Beyond Einstein

Beyond Einstein

The Cosmic Quest for the Theory of the Universe

Michio Kaku
and Jennifer Thompson

Revised and Updated

ANCHOR BOOKS
A DIVISION OF RANDOM HOUSE, INC.
NEW YORK

First Anchor Books Edition, October 1995

 Published in the United States by Anchor Books, a division of Random House, Inc., New York, and simultaneously in Canada by Random House of Canada Limited, Toronto. Originally published in hardcover in the United States by Bantam Books in 1987. The Anchor Books edition is published by arrangement with Bantam Books.

Library of Congress Cataloging-in-Publication Data

Kaku, Michio.
Beyond Einstein: the cosmic quest for the theory of the universe / by Michio Kaku and Jennifer Trainer Thompson.—Rev. and updated.
p. cm.
Includes bibliographical references and index
1. Unified field theories. 2. Superstring theories. 3. Supersymmetry.
4. Cosmology. I. Thompson, Jennifer Trainer. II. Title.
QC173.7.K35 1995
530.1—dc20 95-1815
CIP

ISBN 0-385-47781-3

www.anchorbooks.com

Printed in the United States of America
19 18

Contents

Introduction

The idea for this book dates back to the mid-1950s, when Michio was a child growing up in California and first heard about the unified field theory.

Michio was in fourth grade when he read about the death of a great scientist, Albert Einstein. He learned that Einstein had discovered many great things in his lifetime that made him world famous, but that he had died before he could finish his greatest work. Michio was fascinated by the story.

If the man was that great, the boy reasoned, then his unfinished project must have been wonderful—the crowning achievement in his illustrious career.

Curious, Michio combed the Palo Alto libraries to discover more about this unified field theory, but he couldn't find any books or articles on the subject. There were a few college texts on quantum mechanics, but at age eight Michio found them largely incomprehensible. Moreover, they didn't make even a passing reference to the unified field theory.

So Michio went to his teachers, who had no answers for him. Even physicists whom he later met would shrug their shoulders when he mentioned Einstein's last theory. Most physicists felt that it was premature, or downright presumptuous, to believe that man could unite the four forces in the universe.

Years later, while working on the string theory (which was being proposed as a theory of strong interactions), Michio too grew cynical, believing that perhaps the search for the unified field theory was a wild goose chase after all. No one took physicists John Schwarz and Joel Scherk seriously in the 1970s when they proclaimed that perhaps a sophisticated version of this string theory was the fabled unified field theory that had eluded Einstein and other physicists.

Finally, in 1984, a dramatic theoretical breakthrough was made that seemed to clinch it. ''Superstrings,'' as Schwarz and Scherk had predicted years earlier, seemed the best (and only) candidate for the unified field theory.

Although the details of the theory are still being worked out, it was clear that this discovery was going to shake the world of physics. Michio and Jennifer Thompson had already coauthored a book, *Nuclear Power: Both Sides,* and it seemed natural to team up again and answer the question that had fascinated Michio thirty years earlier: ''What is the unified field theory?''

Together we sought to produce a book that would serve as a guide for the curious layman. We wanted to write a book that covered the ''superstring revolution'' with the insight and scope that often only an insider can provide, and to present the subject in a lively, informative manner. We felt that our combined experience—as a theoretical physicist and as a writer—worked well in this regard.

We also wanted to provide a comprehensive glimpse of the world of physics, presenting the superstring theory in the context of the last three hundred years of science. Many books address one aspect of modern physics—be it relativity, quantum mechanics, or cosmology —but neglect the larger sweep of physics. *Beyond Einstein* is different; instead of dwelling on isolated areas of research, we focus on the entire scope of physics, pointing out where each particular theory fits into the larger picture. What does the unified field theory have to do with quantum mechanics? How does Newton's theory of gravity apply to the superstring theory? These are a few of the questions answered in *Beyond Einstein.*

In this book, we have stressed how superstring theory gives a unified description of *matter.* We have focused on the diverse properties of the subatomic particles, such as the quarks, leptons, Yang-Mills particles, gluons, and others, and how they can be viewed as different vibrations of the superstring. In a companion volume, *Hyperspace,* Michio focuses instead on the properties of *space and time,* especially the possibility of parallel universes, time warps, and the tenth dimension.

We're excited about the new breakthroughs in physics, and we hope we've written a book that is both authoritative and interesting —in short, one that Michio would like to have read when he was young.

New York, N.Y. — Michio Kaku
Williamstown, Ma. — Jennifer Thompson

I

A Theory of the Universe

1

Superstrings: A Theory of Everything?

A NEW THEORY is rocking the foundations of modern physics, rapidly overturning cherished but obsolete notions about our universe and replacing them with new mathematics of breathtaking beauty and elegance. Although there are still some unresolved questions concerning this theory, the excitement among physicists is palpable; throughout the world, leading physicists are proclaiming that we are witnessing the genesis of a new physics.

This theory is called "superstrings," and a series of astonishing breakthroughs in physics within the last decade have culminated in its development, indicating that perhaps we are finally closing in on the unified field theory: a comprehensive, mathematical framework that would unite all known forces of the universe.

Advocates of superstrings even claim that the theory could be the ultimate "theory of the universe."

Although physicists are usually cautious in their approach to new ideas, Princeton physicist Edward Witten has claimed that the superstring theory will dominate the world of physics for the next fifty years. "Superstring theory is a miracle, through and through," he said recently. At one physics conference, he astonished his audience by declaring that we may be witnessing a revolution in physics as great as the birth of the quantum theory. He added, "It's probably going to lead to a new understanding of what space and time really are, the most dramatic [understanding] since general relativity.''[1]

Even *Science* magazine, always careful not to exaggerate the claims of scientists, compared the birth of the superstring theory to the discovery of the Holy Grail. This revolution, *Science* magazine claimed, may be ''no less profound than the transition from real numbers to complex numbers in mathematics.''[2]

Two of the theory's creators, John Schwarz of the California Institute of Technology and Michael Green of Queen Mary College in London, call it—a bit puckishly—a Theory of Everything (TOE).[3]

At the heart of this excitement is the realization that superstrings may provide a comprehensive theory that can explain *all* known physical phenomena—everything from the motion of galaxies down to the dynamics within the nucleus of the atom. The theory even makes startling predictions concerning the origin of the universe, the beginning of time, and the existence of multidimensional universes.

To a physicist, it is an intoxicating notion that the vast storehouse of information of our physical universe, painfully accumulated over several thousand years of careful investigation, can be summarized in one theory.

For example, German physicists have compiled an encyclopedia, the *Handbuch der Physik,* an exhaustive work that summarized the world's knowledge of physics. The *Handbuch,* which physically occupies an entire bookshelf of a library, represented the pinnacle of scientific learning. If the superstring theory is correct, then all the information contained in this encyclopedia can be derived (in principle) from *a single equation.*

Physicists are particularly excited about the superstring theory because it forces us to revise our understanding of the nature of matter. Since the time of the Greeks, scientists have assumed that the building blocks of the universe were tiny point particles. Democritus coined the word *atomos* to describe these ultimate, indestructible units of matter.

The superstring theory, however, assumes that the ultimate building blocks of nature consist of tiny vibrating strings. If correct, this means that the protons and neutrons in all matter, everything from our bodies to the farthest star, are ultimately made up of strings. Nobody has seen these strings because they are much too small to be observed. (They are about *100 billion billion* times smaller than a proton.) According to the superstring theory, our world only appears to be made of point particles, because our measuring devices are too crude to see these tiny strings.

At first it seems strange that such a simple concept—replacing point particles with strings—can explain the rich diversity of particles and forces (which are created by the exchange of particles) in nature. The superstring theory, however, is so elegant and comprehensive that it is able to explain simply why there can be billions upon billions of different types of particles and substances in the universe, each with astonishingly diverse characteristics.

The superstring theory can produce a coherent and all-inclusive picture of nature similar to the way a violin string can be used to "unite" all the musical tones and rules of harmony. Historically, the laws of music were formulated only after thousands of years of trial-and-error investigation of different musical sounds. Today, these diverse rules can be derived easily from a single picture—that is, a string that can resonate with different frequencies, each one creating a separate tone of the musical scale. The tones created by the vibrating string, such as C or B flat, are not in themselves any more fundamental than any other tone. What is fundamental, however, is the fact that a single concept, vibrating strings, can explain the laws of harmony.

Knowing the physics of a violin string, therefore, gives us a comprehensive theory of musical tones and allows us to predict new harmonies and chords. Similarly, in the superstring theory, the fundamental forces and various particles found in nature are nothing more than different modes of vibrating strings. The gravitational interaction, for example, is caused by the lowest vibratory mode of a circular string (a loop). Higher excitations of the string create different forms of matter. From the point of view of the superstring theory, no force or particle is more fundamental than any other. All particles are just different vibratory resonances of vibrating strings. Thus, a single framework—the superstring theory—can in principle explain why the universe is populated with such a rich diversity of particles and atoms.

The answer to the ancient question "What is matter?" is simply that matter consists of particles that are different modes of vibration of the string, such as the note G or F. The "music" created by the string is matter itself.

But the fundamental reason why the world's physicists are so excited by this new theory is that it appears to solve perhaps the most important scientific problem of the century: namely, how to unite the four forces of nature into one comprehensive theory. At the center of this upheaval is the realization that the four fundamental forces governing our universe are actually different manifestations of a single unifying force, governed by the superstring.

Four Forces

A force is anything that can move an object. Magnetism, for example, is a force because it can make a compass needle spin. Electricity is a force because it can make our hair stand on end. Over the last two thousand years, we gradually have realized that there are four fundamental forces: gravity, electromagnetism (light), and two types of nuclear forces, the weak and the strong. (Other forces identified by the ancients, such as fire and wind, can be explained in terms of the four forces.) One of the great scientific puzzles of our universe, however, has been why these four forces seemed so different. For the past fifty years, physicists have grappled with the problem of uniting them into a coherent picture.

To help you appreciate the excitement that the superstring theory is generating among physicists, we will take a minute to describe each force and show just how dissimilar they are.

Gravity is an attractive force that binds together the solar system, keeps the earth and the planets in their orbits, and prevents the stars from exploding. In our universe, gravity is the dominant force that extends trillions upon trillions of miles, out to the farthest stars; this force, which causes an apple to fall to the ground and keeps our feet on the floor, is the same force that guides the galaxies in their motions throughout the universe.

The electromagnetic force holds together the atom. It makes the electrons (with negative charge) orbit around the positively charged nucleus of the atom. Because the electromagnetic force determines the structure of the orbits of the electrons, it also governs the laws of chemistry.

On the earth, the electromagnetic force is often strong enough to overpower gravity. By rubbing a comb, for example, it is possible to pick up scraps of paper from a table. The electromagnetic force counteracts the downward force of gravity and dominates the other forces down to .000000000001 inch (roughly the size of a nucleus).

(Perhaps the most familiar form of the electromagnetic force is light. When the atom is disturbed, the motion of the electrons around the nucleus becomes irregular, and the electrons emit light and other forms of radiation. This is the purest form of electromagnetic radiation, in the form of X rays, radar, microwave, or light. Radio and television are simply different forms of the electromagnetic force.)

Within the nucleus of the atom, the electromagnetic force is overpowered by the weak and strong (nuclear) forces. The strong force, for example, is responsible for binding together the protons and neutrons in the nucleus. In any nucleus, all the protons are positively charged. Left to themselves, their repulsive electric force would tear apart the nucleus. The strong force, therefore, overcomes the repulsive force between the protons. Roughly speaking, only a few elements can maintain the delicate balance between the strong force (which tends to hold the nucleus together) and the repulsive electric force (which tends to rip apart the nucleus), which helps to explain why there are only about one hundred known elements in nature. Should a nucleus contain more than about a hundred protons, even the strong nuclear force would have difficulty containing the repulsive electric force between them.

When the strong nuclear force is unleashed, the effect can be catastrophic. For example, when the uranium nucleus in an atomic bomb is split deliberately, the enormous energies locked within the nucleus are released explosively in the form of a nuclear detonation. Pound for pound, a nuclear bomb releases over a million times the energy contained in dynamite. Indeed, the strong force can yield significantly more energy than a chemical explosive, which is governed by the electromagnetic force.

The strong force also explains the reason why stars shine. A star is basically a huge nuclear furnace in which the strong force within the nucleus is unleashed. If the sun's energy, for example, were created by burning coal instead of nuclear fuel, only a minuscule fraction of the sun's light would be produced. The sun would rapidly fizzle and turn into a cinder. Without sunlight, the earth would turn cold and life on it would eventually die. Without the strong force, therefore, the stars would not shine, there would be no sun, and life on earth would be impossible.

If the strong force were the only force at work inside the nucleus, then most nuclei would be stable. However, we know from experience that certain nuclei (such as uranium, with ninety-two protons) are so massive that they automatically break apart, releasing smaller fragments and debris, which we call radioactivity. In these elements the nucleus is unstable and disintegrates. Therefore, yet another, weaker force must be at work, one that governs radioactivity and is responsible for the disintegration of very heavy nuclei. This is the weak force.

The weak force is so fleeting and ephemeral that we do not experience it directly in our lives. However, we feel its indirect effects. When a Geiger counter is placed next to a piece of uranium, the clicks that we hear measure the radioactivity of the nuclei, which is caused by the weak force. The energy released by the weak force can also be used to create heat. For example, the intense heat found in the interior of the earth is partially caused by the decay of radioactive elements deep in the earth's core. This tremendous heat, in turn, can erupt in volcanic fury if it reaches the earth's surface. Similarly, the heat released by the core of a nuclear power plant, which can generate enough electricity to light up a city, also is caused by the weak force (as well as the strong force).

Without these four forces, life would be unimaginable: The atoms of our bodies would disintegrate, the sun would burst, and the atomic fires lighting the stars and galaxy would be snuffed out. The idea of forces, therefore, is an old and familiar one, dating back at least to Isaac Newton. What is new is the idea that these forces are nothing but different manifestations of a single force.

Everyday experience demonstrates the fact that an object can manifest itself in a variety of forms. Take a glass of water and heat it until it boils and turns into steam. Water, normally a liquid, can turn into steam, a gas, with properties quite unlike any liquid, but it is still water. Now freeze the glass of water into ice. By withdrawing heat, we can transform this liquid into a solid. But it is still water—

the same substance—merely turned into a new form under certain circumstances.

Another, more dramatic example is the fact that a rock can turn into light. Under specific conditions, a piece of rock can turn into vast quantities of energy, especially if that rock is uranium, and the energy manifests itself in an atomic bomb. Matter, then, can manifest itself in two forms—either as a material object (uranium) or as energy (radiation).

In much the same way, scientists have realized over the past hundred years that electricity and magnetism are manifestations of the same force. Only within the last twenty-five years, however, have scientists understood that even the weak force can be treated as a manifestation of the same force. The Nobel Prize in 1979 was awarded to three physicists (Steven Weinberg, Sheldon Glashow, and Abdus Salam) who showed how to unite the weak and the electromagnetic forces into one force, called the "electro-weak" force. Similarly, physicists now believe that another theory (called the GUT, or "grand unified theory") may unite the electro-weak force with the strong interactions.

But the final force—gravity—has long eluded physicists. In fact, gravity is so unlike the other forces that, for the past sixty years, scientists have despaired of uniting it with the others. Although quantum mechanics spectacularly united the other three forces, it failed dismally when applied to gravity.

The Missing Link

In the twentieth century, two great theories have towered above all others: quantum mechanics, with its resounding success in explaining the three subatomic forces, and Einstein's theory of gravity, called general relativity. In some sense, these two theories are opposites: While quantum mechanics is devoted to the world of the very small—such as atoms, molecules, protons, and neutrons—relativity governs the physics of the very large, on the cosmic scale of stars and galaxies.

To physicists, one of the great puzzles of this century has been that these two theories, from which we can in principle derive the sum total of human knowledge of our physical universe, should be so incompatible. In fact, merging quantum mechanics with general relativity has defied all attempts by the world's greatest minds in this century. Even Albert Einstein spent the last three decades of his life on a futile search for a unifying theory that would include gravity and light.

Each of these two theories, in its particular domain, has scored spectacular successes. Quantum mechanics, for example, has no rival in explaining the secrets of the atom. Quantum mechanics has unraveled the secrets of nuclear physics, unleashed the power of the hydrogen bomb, and explained the workings of everything from transistors to lasers. In fact, the theory is so powerful that, if we had enough time, we could predict all the properties of the chemical elements by computer, without ever having to enter a laboratory. However, although quantum mechanics has been undeniably successful in explaining the world of the atom, the theory fails when trying to describe the gravitational force.

On the other hand, general relativity has scored brilliant successes in its own domain: the cosmic scale of galaxies. The black hole, which physicists believe is the ultimate state of a massive, dying star, is a well-known prediction of general relativity. General relativity also predicts that the universe originally started in a Big Bang that sent the galaxies hurtling away from one another at enormous speeds. The theory of general relativity, however, cannot explain the behavior of atoms and molecules.

So, physicists were faced with two distinct theories, each employing a different set of mathematics, each making astonishingly accurate predictions within its own realm, each profoundly separate and distinct.

It's as if nature created someone with two hands, with the right hand looking entirely different and functioning totally independently from the left hand. For physicists, who believe that nature ultimately should be simple and elegant, it was a puzzle; they could not believe that nature could function in such a bizarre fashion.

This is where superstrings enter the picture, for they may solve the problem of how to embrace these two great theories. In fact, both halves—quantum mechanics and relativity—are *necessary* to make the superstring theory work. Superstrings are the first and *only* mathematical framework in which a quantum theory of gravity makes sense. It's as if scientists for the past six decades were trying to assemble a cosmic jigsaw puzzle and suddenly noticed that the missing piece were superstrings.

Stranger Than Science Fiction

Ordinarily, scientists are conservative. They are slow to accept new theories, especially those that make predictions that are the least bit strange. The superstring theory, however, makes some of the wildest predictions of any theory ever proposed. Any theory that has the ability to condense the essence of so much physics into one equation will have profound physical consequences, and this theory is no exception.

(In 1958, the great quantum physicist Niels Bohr attended a talk given by physicist Wolfgang Pauli. At the end of the talk, which the audience received unfavorably, Bohr remarked, "We all agree that your theory is crazy. The question which divides us is whether it is crazy enough." Superstring theory, because of its bizarre predictions, is certainly "crazy enough.")

Although these predictions are discussed fully in ensuing chapters, a few of them are touched on here, to provide a glimpse of what people mean when they say that superstrings suddenly make real physics look stranger than science fiction.

MULTIDIMENSIONAL UNIVERSES

In the 1920s, Einstein's general theory of relativity provided the best explanation of how our universe began. According to Einstein's theory, the universe was born approximately 10 to 20 billion years ago in a gigantic explosion called the Big Bang. All the matter in the universe, including the stars, galaxies, and planets, was originally concentrated in one superdense ball, which exploded violently, creating our current expanding universe. This theory explains the observed fact that all the stars and galaxies are currently moving away from the earth (propelled by the force of the Big Bang).

However, there were many gaps in Einstein's theory. Why did the universe explode? What happened before the Big Bang? Theologians as well as scientists have for years realized the incompleteness of the Big Bang theory, because it fails to explain the origin and nature of the Big Bang itself.

Incredibly, the superstring theory predicts what happened before the Big Bang. According to superstrings, the universe originally existed in ten dimensions, not the four dimensions (three space dimensions and one time dimension) of today. However, because the universe was unstable in ten dimensions, it "cracked" into two pieces, with a small, four-dimensional universe peeling off from the rest of the universe. By analogy, imagine a soap bubble that is vibrating slowly. If the vibrations become strong enough, the soap bubble becomes unstable and fissions into two or more smaller soap bubbles. Imagine that the original soap bubble represents the ten-dimensional universe, and that one of the smaller soap bubbles represents our universe.

If this theory is true, it means that our universe actually has a "sister universe" that coexists with our universe. It also means that the original fissioning of our universe was so violent that it created the explosion that we know as the Big Bang. The superstring theory, therefore, explains the Big Bang as a by-product of a much more violent transition, the cracking of the ten-dimensional universe into two pieces.

You do not have to worry, however, that one day as you are walking down the street you will "fall" into another other-dimensional universe as if in a science fiction novel. According to the superstring theory, the other multidimensional universe has shrunk to such an incredibly small size (about 100 billion billion times smaller than the nucleus of an atom) that it can never be reached by humans. Thus, it becomes an academic question what higher dimensions look like. In this sense, the prospect of traveling between higher dimensions was possible only at the origin of the universe, when the universe was ten-dimensional and interdimensional travel was physically possible.

DARK MATTER

In addition to multidimensional spaces, science fiction writers sometimes spice up their novels with talk of "dark matter," a mysterious form of matter with properties unlike any found in the universe.

Dark matter was predicted in the past, but wherever scientists trained their telescopes and instruments in the heavens, they found only the hundred or so familiar chemical elements existing on the earth. Even stars in the farthest reaches of the universe are made of ordinary hydrogen, helium, oxygen, carbon, et cetera. On one hand, this was reassuring; we knew that wherever we traveled in outer space, our rocket ships would encounter only the chemical elements found on the earth. On the other hand, it was a bit disappointing knowing that there would be no surprises in outer space.

The superstring theory might possibly change that, for the process of fissioning from a ten-dimensional universe down to smaller universes probably created a new form of matter. This dark matter has weight, like all matter, but is invisible (hence the name). Dark matter is also tasteless and has no smell. Even our most sensitive instruments cannot detect its presence. If you could hold this dark matter in your hand, it would feel heavy, but it would otherwise be undetectable. In fact, the only way to detect dark matter is by its weight: it has no other known interaction with other forms of matter.

Dark matter also may help to explain one of the puzzles of cosmology. If there is sufficient matter in the universe, then the gravitational attraction of the galaxies should slow down its expansion and even possibly reverse it, causing the universe to collapse. However, there is conflicting data as to whether there is enough matter in the universe to cause this reversal and eventual collapse. Astronomers who have tried to calculate the total amount of matter in the visible universe find that there is simply not enough matter in stars and galaxies to cause the universe to collapse. However, other calculations (based on calculating the red shifts and luminosities of stars) indicate that the universe might collapse. This is called the "missing mass" problem.

If the superstring theory is correct, then it may explain why astronomers fail to see this form of matter in their telescopes and instruments. Moreover, if the theory of dark matter is correct, dark matter may pervade the universe. (Indeed, there may be more dark matter than ordinary matter.) In this regard, the superstring theory not only clarifies what happened before the Big Bang but predicts what may happen at the death of the universe.

Super Skeptics

Of course, any theory that makes claims of this magnitude—to replace point particles with strings and a four-dimensional universe with a ten-dimensional one—invites skepticism. Although the superstring theory opens up a vista of mathematics that has startled even the mathematicians and has excited physicists from around the world, it may take years or even decades before we can build machines powerful enough to test the theory conclusively. Meanwhile, until there is irrefutable experimental proof, skeptics remain unconvinced of the superstring theory, despite its beauty, elegance, and uniqueness.

"Years of intense effort," complained Harvard physicist Sheldon Glashow, "by dozens of the best and the brightest have yielded not one verifiable prediction, nor should any soon be expected."[4]

World-renowned Dutch physicist Gerard 't Hooft, speaking at the Argonne National Laboratory outside Chicago, went so far as to compare the fanfare surrounding superstrings to "American television commercials"[5]—all advertisement and very little substance.

Indeed, as Princeton physicist Freeman Dyson once cautioned, referring in general to the search for a single mathematical model that would describe the unification of all four forces: "The ground of physics is littered with the corpses of unified theories."[6]

But superstrings' defenders point out that, although a decisive experiment that could prove the theory may be years away, there are no experiments that contradict the theory. No other theory can make that claim.

Indeed, the theory has no rival: There is no other way at the present time to marry the quantum and relativity theories consistently. Some physicists are skeptical of new attempts to find a unifying theory because so many attempts failed in the past, but these attempts failed because they could not unite gravity with quantum mechanics. The superstring theory, however, seems to accomplish this; it does not suffer from the disease that killed off its predecessors. Because of this, the superstring theory is by far the most promising candidate for a true unification of all forces.

The SSC—Largest Scientific Machine in History

The world of physics, which is closing in on a unified description of the weak, the electromagnetic, the strong, and possibly the gravitational interactions, has spawned efforts to create powerful machines to test certain aspects of these theories. These theories are not matters of idle speculation but are the focal point of intense international interest.

For much of the 1980s, the U.S. government was committed to spending billions to build a colossal "atom smasher" or particle accelerator to probe deep into the atom's nucleus. This machine, called the Superconducting Super Collider (SSC), would have been the largest scientific machine ever built; however, the project was canceled in 1993.

The primary mission of the SSC was to find new interactions and test the predictions of these unified theories, such as the electroweak theory, and possibly probe the fringes of the GUT and the superstring theory. This powerful machine would have focused on various aspects of the search for this fabled unification. Devouring enough energy to power a large metropolis, the SSC would have accelerated particles to trillions of electron volts in order to smash other subatomic particles. Physicists were hoping that locked deep within the nucleus of the atom was the crucial data necessary to verify some aspects of these theories.

The SSC, which would have dominated experimental high-energy physics into the next century, would still not have been large enough to test fully the consequences of the GUT theory, which unites the strong force with the electro-weak force, or the more ambitious superstring theory, which unites all known forces. Testing the predictions of both these theories would require machines vastly larger than the SSC. However, the SSC may have been able to probe the periphery of these theories and help us indirectly to verify or disprove various predictions of these theories.

Experimentally, because the energies needed to probe the GUT and superstring theories are so fabulously large, the ultimate verification may come from the field of cosmology (the study of the origin of the universe). In fact, the energy scale in which this unification takes place can be found only at the beginning of time. In this sense solving the puzzle of the unified field theory may well solve the riddle of the origin of the universe.

But we are getting ahead of our story. Before one can build a house, one must first lay a foundation. So, too, in physics: Before we can explore in detail how the superstring theory unifies all forces, we must first answer some basic questions, such as: What is relativity? What is matter? Where did the idea of unification originate? These questions are the focus of the following two chapters.

2

The Quest for Unification

HISTORICALLY, science has developed rather disjointedly. The great contributions of Isaac Newton, for example, who computed the motions of the planets with his theory of gravitation, differ significantly from the works of Werner Heisenberg and Erwin Schrödinger, who revealed the workings of the atom with their quantum mechanics. Moreover, the mathematics and principles required for quantum mechanics seem dissimilar to Einstein's general theory of relativity, which describes space warps, black holes, and the Big Bang.

With developments in the unified field theory, however, it now becomes possible to assemble these disjointed pieces and view the whole as more than just the sum of its parts. Although the quest for unification is a recent one, with most of the pioneering work done in the past twenty years, in hindsight it is possible to reanalyze many of the great discoveries in science in terms of the coherent concept of unification.

Due to the momentum created by the unified field theory, the history of science is slowly being rewritten—beginning with the man who practically invented physics, Isaac Newton, and his discovery of the universal law of gravitation, easily the most significant scientific development in several millennia of human history.

Uniting the Heavens and the Earth

Newton lived in the late 1600s, when the church and scholars of the day believed in two distinct types of laws. The laws governing the heavens were perfect and harmonious, while mortals on earth lived under physical laws that were coarse and vulgar.

Anyone who insisted that the moon wasn't a perfect, polished sphere, or that the earth revolved around the sun, could be put to death by the church. Giordano Bruno was burned at the stake in 1600 in Rome for speculating that our sun was just another star and concluding that "there are then innumerable suns, and an infinite number of earths revolve around those suns. . . ."[1] A few decades later, the great astronomer and physicist Galileo Galilei had to recant, on pain of death, his heretical statements that the earth moved around the sun. (Even as he was forced to repudiate his scientific findings at his trial, he is said to have muttered under his breath, "But the earth does move!")

All this began to change when Isaac Newton, a twenty-three-year-old student, was sent home from Cambridge University because the dreaded Black Plague was sweeping the land and had closed down most of the universities and other institutions in Europe. With plenty of time on his hands, Newton observed the motion of objects that fall to the earth and then, in a stroke of brilliance, conceived of his famous theory, which governs the path of all falling objects.

Newton was led to his theory by asking himself such revolutionary questions as: Does the moon also fall?

According to the church, the moon stayed in the sky because it obeyed heavenly laws that were beyond the reach of earthly laws that forced objects to fall to the ground. Newton's revolutionary observation was to extend the law of gravitation into the heavens itself. An immediate conclusion of this heretical idea was that the moon was a satellite of the earth, held in the sky not by the motion of imaginary celestial spheres but by the laws of his gravitation theory.

Perhaps, Newton thought, the moon is continually falling, guided by the same laws that make a rock fall to the earth, but the moon never crashes to the earth because the earth's receding curvature cancels the falling motion. In his masterpiece, *Principia,* Newton wrote down the laws that govern the motion of satellites orbiting the earth and planets orbiting the sun.

Newton drew a simple picture that explained this idea of the falling moon being an earth satellite. Imagine standing on a high mountaintop and throwing a rock, which eventually falls to the earth. The faster you throw the rock, the farther it goes before it falls to the earth. In fact, argued Newton, if the rock were thrown with sufficient velocity, it would circle the earth and hit you in the back of your head. Like a rock circling the earth, the moon is simply a satellite continually falling to the earth.

This elegant picture conceived by Newton predated the launching of artificial satellites by three centuries. Today, the stunning achievements of our space probes, which have landed on Mars and flown past Uranus and Neptune, owe their success to the laws written down by Newton in the late 1600s.

In a rapid series of insights, Newton discovered that his equations allowed him, in principle, to estimate roughly the distance from the earth to the moon and the distance from the earth to the sun. While the church was still teaching that the earth stood still in the heavens, Isaac Newton was calculating the basic dimensions of the solar system.

In retrospect, we can appreciate Newton's discovery of the law of gravitation as the first "unification" in the history of science, uniting the laws of heaven and earth. The same force of gravitation, which acts instantaneously between any two bodies on the earth, linked the destiny of humans with the stars. After Newton's discovery, the motion of the entire solar system could be calculated with almost perfect accuracy.

Furthermore, Newton's diagram showing how even terrestrial rocks can orbit the earth without needing celestial spheres demonstrated that he was able to isolate the essential principles of his theory pictorially. Interestingly, all the great breakthroughs in science, especially those showing the unification of forces, can be displayed graphically. Although the mathematics may be obscure and tedious, the essence of unification is always pictorially quite simple.

Maxwell's Discovery

The next great leap in our understanding of unification—that of electricity and magnetism—took place two hundred years later, in the mid-1860s, during the American Civil War. While the United States was thrown into chaos by that devastating war, across the Atlantic the world of science was also in a period of great turmoil. Experiments being performed in Europe pointed to the unmistakable fact that magnetism, under certain circumstances, can turn into an electric field, and vice versa.

For centuries it was thought that magnetism, the force that guides the compass needles of navigators while at sea, and electricity, the force that creates everything from lightning bolts to the shock upon touching a doorknob after walking across a carpet, were distinct forces. However, by the mid-1800s, this rigid separation was falling apart as scientists realized that vibrating electric fields could create magnetic ones, and vice versa.

This effect can be demonstrated easily. For example, simply by shoving a bar magnet into a coil of wire we can generate a small electric current within the wire. Thus, a changing magnetic field has created an electric field. Similarly, we can reverse this demonstration by running an electric current through this coil of wire, thereby producing a magnetic field around the coil. Thus, a changing electric field has now created a magnetic field.

This same principle—that changing electric fields can produce magnetic fields and vice versa—is the reason why we have electricity in our homes. In a hydroelectric plant, water falling over a dam rotates a huge wheel connected to a turbine. The turbine contains large wire coils that spin rapidly in a magnetic field. Electricity is created by the spinning motion of these coils as they move in the magnetic field. This electricity, in turn, is sent over hundreds of miles of wires into our homes. Thus, a changing magnetic field (created by the dam) is transformed into an electric field (which brings electricity into our homes through our wall sockets).

In 1860, however, this effect was understood poorly. An obscure thirty-year-old Scottish physicist at Cambridge University, James Clerk Maxwell, challenged the prevailing thinking of the day and claimed that electricity and magnetism were not distinct forces but two sides of the same coin. In fact, he made the most astonishing discovery of the century when he found that this observation could unlock the secret to the most mysterious phenomenon of all: light itself.

Electric and magnetic fields, Maxwell knew, could be visualized as ''force fields'' that permeate all space. They can be represented by an infinite array of ''arrows'' emanating smoothly from an electric charge. For example, the force fields created by a bar magnet reach into space like a spiderweb and can ensnare nearby metallic objects.

Maxwell went further than this, however, and argued that it might be possible for electric and magnetic fields to vibrate together in precise synchronization, so that they generated a wave that could travel by itself in space without assistance.

One can visualize the following scenario: What would happen if a vibrating magnetic field created an electric field, which in turn vibrated and created yet another magnetic field, which in turn vibrated and created still another electric field, et cetera? Wouldn't such an infinite chain of vibrating electric and magnetic fields travel by itself, much like a wave?

As with Newton's laws of gravitation, the essence of the idea is simple and pictorial. Think, for example, of a long line of dominoes. Tipping over the first domino, of course, will trigger a cascading wave of falling dominoes. Let's say, however, that this line of dominoes consists of two types, colored black and white, and that these colored dominoes alternate along the line. If we remove the black dominoes, leaving only the white ones in the line, then a wave can no longer travel. We need both the white and the black dominoes in order to have a traveling wave. In summary, it is the interplay of white and black dominoes, with each one tipping over the next, that makes possible the wave of falling dominoes.

Similarly, Maxwell discovered that the interplay between vibrating magnetic and electric fields created the wave. He found that electric or magnetic fields alone could not create this wavelike motion, similar to the analogy with only black or white dominoes. Only the delicate interplay between electric and magnetic fields could produce this wave.

To most physicists, however, the idea seemed preposterous because there was no "ether" to conduct these waves. These fields were "disembodied" and moved by themselves, without a conducting medium.

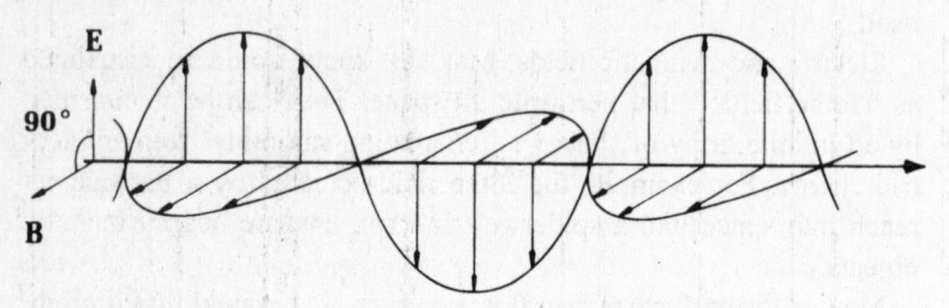

According to Maxwell's theory, light consists of electric fields (E) and magnetic fields (B) that oscillate in unison. Here the electric fields vibrate vertically while the magnetic fields vibrate horizontally.

Maxwell was undaunted, however. By calculating with his equations, he found that he was able to derive a specific number for the speed of this wave. Much to his astonishment, he found it to be the speed of light. The conclusion was inescapable: light was revealed as nothing but a chain of electric fields turning into magnetic fields. Quite by accident, Maxwell found that his equations unraveled the nature of light as an *electromagnetic* wave. Therefore, he was the first to discover a genuine unified field theory.

This was a fantastic discovery, ranking in importance alongside Newton's discovery of the universal law of gravitation. In 1889, ten years after Maxwell's death, Heinrich Hertz experimentally confirmed Maxwell's theories. In a dramatic demonstration, Hertz generated an electric spark and was able to create an electromagnetic wave that was detected unmistakably over large distances. Just as Maxwell had predicted, Hertz proved that these waves traveled by themselves, without "ether." Eventually, Hertz's crude experiment evolved into the vast industry we call "radio."

Due to Maxwell's pathbreaking work, from then on light would be known as the electromagnetic force, created by the vibration of electric and magnetic fields rapidly turning into each other. Radar, ultraviolet rays, infrared rays, radio, microwaves, television, and X rays are nothing but different forms taken by the electromagnetic wave. (For example, when you tune in to your favorite radio station on, say, 99.5 on the dial, the electric and magnetic fields contained in that radio wave are turning into each other at the rate of 99.5 million times per second.)

Unfortunately, Maxwell died soon after proposing this theory; he did not live long enough to probe the peculiarities of his creation. A perceptive physicist, however, could have noticed even in the 1860s that Maxwell's equations necessarily required bizarre distortions of distance and time. His equations were fundamentally different from Newton's theory because of the way space and time were described. To Newton, the pulse of time beat uniformly throughout the cosmos. A clock on the earth beat at the same rate as a clock on the moon. Maxwell's equations, however, predicted that, under certain circumstances, clocks could slow down.

Scientists failed to realize that Maxwell's theory predicted that a clock placed on a moving rocket ship should beat slower than a clock placed on the earth. At first, this sounds totally absurd. After all, the uniformity of the passage of time was one of the foundations of the Newtonian system. Nonetheless, Maxwell's equations required this strange distortion of time.

For half a century, however, scientists overlooked this strange prediction of Maxwell's equations. It wasn't until 1905 that a physicist finally understood this profound distortion of space and time that was built into Maxwell's theory. The physicist was Albert Einstein, and the theory he created was the special theory of relativity, which would change the course of human history.

An Unemployed Revolutionary

In his lifetime, Einstein proposed many ideas that would revolutionize the way we view the universe. However, if we summarized his body of work, we could categorize it into three broad classes of theories: special relativity, general relativity, and his unfinished unified field theory, which was to have been his crowning scientific achievement.

He proposed his first great theory—the special theory of relativity—in 1905, when he was only twenty-six years old. For a man who would have such an impact on the world of science, his origins were humble.

In 1900, the future world-renowned physicist found himself out of a job and out of luck. While better-known physicists lectured at the great universities, Einstein had been rejected for a teaching position by various universities. Having just completed his studies at the Zurich Polytechnic, he was struggling to survive by part-time tutoring. His father, concerned about his son's depression, wrote, "My son is deeply unhappy with his current state of unemployment. Day by day the feeling grows in him that his career is off the track . . . the awareness weighs on him that he is a burden to us, people of small means."[2]

In 1902, through the recommendation of a friend, he landed a humble job with the patent office in Bern, Switzerland, to support his wife and child. Although Einstein was overqualified for this job, in hindsight it served his purposes surprisingly well.

First, the patent office was a quiet refuge that gave Einstein plenty of time to tinker with a new theory of space and time that he was investigating. Second, his work at the patent office required him to isolate the key ideas from inventors' often vaguely worded proposals. This taught him, like Newton and Maxwell before him, how to think in terms of physical pictures and to zero in unerringly on the fundamental ideas that make a theory work.

At the patent office, Einstein returned to a question that had bothered him even as a child: What would a beam of light look like if he could race next to it at the speed of light? At first, one would suspect that the wave of light would be frozen in time, so that one could actually see the stationary waves of electric and magnetic fields.

But when Einstein learned Maxwell's equations at the Polytechnic, he was surprised to find that they do *not* admit stationary waves as solutions. In fact, Maxwell's equations predict that light must travel at the same speed, regardless of how hard one tries to catch up with it. Even if a person traveled at enormous velocities, a light beam would precede him at the same speed. Light waves can never be seen at rest.

At first, this seems deceptively simple. According to Maxwell's equations, a scientist on the earth and a scientist on a speeding rocket ship will measure the same velocity for a light beam. Perhaps Maxwell himself, writing in the 1860s, realized this. However, only Einstein grasped the singular importance of this fact, for he realized that this meant we must change our notions of space and time. In 1905, Einstein finally solved the puzzle of Maxwell's theory of light. In the process, he overturned the notions of space and time that had survived for several thousand years.

For the sake of argument, say that the speed of light is 101 miles per hour. It should then be possible for a train traveling at 100 miles per hour to move practically side by side with a light beam. In fact, a scientist on this train should measure the speed of light to be 1 mile per hour (101 mph minus 100 mph). The scientist should be able to study the internal structure of light leisurely, in detail.

According to Maxwell's equations, however, the scientist traveling at 100 miles per hour should measure the velocity of the light beam to be 101 miles per hour, not 1 mile per hour. How is this possible? How can the scientist on the train be fooled into thinking that the beam of light is traveling so fast?

Einstein's solution to this problem was outlandish but correct: he postulated that clocks on the train beat much slower than clocks on the earth and that any measuring sticks on the train shrank in length.

This means that the brain of the scientist on the train slows down relative to the brain of the scientist on earth. As seen from someone on earth, the scientist on the train *should* measure the light beam moving at 1 mile per hour, but in reality the scientist measures the velocity of the light beam to be 101 miles per hour because his brain (and everything inside the train) has slowed down.

The consequences of relativity—that time must slow down and distances must contract for speeding bodies—seem to violate common sense. This is only because common sense deals with occurrences far removed from the speed of light. Humans can walk at about 5 miles per hour—much slower than the speed of light. So, for all intents and purposes, humans act on the intuitive assumption that the speed of light is infinite. Light, which can travel around the earth seven times in one second, from our point of view essentially moves instantaneously.

However, imagine a world in which the speed of light is only 5 miles per hour, the speed of an average stroller. If the speed of light were 5 miles per hour, then it would be "common sense" that time and space undergo vast distortions. For example, cars wouldn't be able to travel more than 5 miles per hour, and those traveling near 5 miles per hour would be flattened, like pancakes. (By a curious effect, however, these shrunken cars would not only look flattened to an observer, but also rotated.) Furthermore, the flattened people in these cars would appear to be almost motionless and frozen in time. (This is because time slows down as the cars speed up.) As these flattened cars slowed down at a traffic light, however, they would gradually expand in length until they assumed their original size, and time would return to normal inside the car.

When Einstein's revolutionary 1905 paper was published, it was largely ignored. In fact, he submitted the paper in order to secure a teaching position at the University of Bern, and the paper was rejected. To the classical Newtonian physicist, schooled in the notion of absolute space and absolute time, Einstein's proposal was perhaps the most extreme solution to the paradoxes of Maxwell's equations. (Only years later, when the experimental evidence pointed out the correctness of Einstein's theory, did the scientific community realize that the idea in the paper contained a stroke of genius.)

Decades later, Einstein would acknowledge the importance of Maxwell's theory to the development of special relativity when he said flatly, "The special theory of relativity owes its origin to Maxwell's equations of the electromagnetic field."[3]

In hindsight, however, we realize that Einstein was able to take Maxwell's theory further than anyone else because he grasped the principle of unification, understanding that there was an *underlying, unifying symmetry** that linked seemingly dissimilar objects such as space and time (as well as matter and energy). Like Newton's seminal discovery that terrestrial and celestial physics could be united by his universal law of gravitation, or Maxwell's discovery of the unity of electricity and magnetism, Einstein's contribution was to unite space and time.

This theory demonstrated that space and time are manifestations

* To a physicist, symmetry has a precise meaning; an equation has symmetry if it remains unchanged when we shuffle or rotate its components. Symmetry has emerged as the most powerful tool by which physicists construct the unified field theory. For more detail, see chapter 7.

of one entity, which scientists call "space-time." However, the theory not only united space and time, it joined the concepts of matter and energy.

At first glance, nothing can be farther apart in appearance than an ugly rock and a brilliant, radiant beam of light. Appearances, however, are deceiving. It was Einstein who first pointed out that, under certain circumstances, even a rock (uranium) can turn into a beam of light (a nuclear detonation). The conversion of matter into energy is carried out by the splitting of the atom, which releases the tremendous energy stored within the nucleus. The essence of relativity lies in Einstein's realization that matter can turn into energy and vice versa.

Space Warps

Although Einstein's theory of special relativity received wide recognition in the years after its proposal, Einstein was not satisfied with the theory. To him, it was still incomplete, with nothing more glaring than the fact that the theory omitted any reference to gravity. Newton's theory of gravity, it appeared, violated the basic principles of special relativity.

Imagine what would happen if the sun disappeared suddenly. How long would it take for the earth to spin wildly out of its orbit? According to Newton's theory, if the sun disappeared, the earth would fly into deep space immediately, leaving the solar system.

To Einstein, this conclusion was unacceptable; nothing, including gravity, could go faster than the speed of light. It should take eight minutes (the time it takes for light from the sun to reach the earth) before the earth wobbled out of its orbit. This, of course, required a new theory of gravity. Newton's theory of gravity *must* be wrong because it made no reference to the speed of light, the ultimate velocity in the universe.

Einstein's solution to this puzzle, which he proposed in 1915, was the theory of general relativity, which explained gravitation as the marriage of space-time and matter-energy. Once again, although the mathematics of the equations were complex, the theory can be summarized by a simple physical picture.

Imagine a trampoline net with a bowling ball placed in the center.

Naturally, the weight of the ball will make the trampoline net sink. Now consider a small marble that is moving along the surface of the curved net. The marble, instead of executing a straight line, will travel in a circular orbit around the depression caused by the ball.

According to Newton, one can imagine an invisible "force" acting between the ball and the marble. According to Einstein, however, a much simpler interpretation is that the warping of the surface of the net by the ball causes the marble to move in circles.

Now imagine that the ball is really our sun, the marble is the earth, and the trampoline net is space-time. We suddenly realize that "gravity" is not a force at all, but the bending of space-time caused by the presence of matter-energy (the sun).

If the ball was suddenly removed from the trampoline net, the vibrations caused by its removal would travel like a wave along the surface of the net. A fraction of a second later, the wave would hit the marble, and the marble's course would be altered. This is the solution, then, to the problem of what would happen if the sun suddenly disappeared. The waves of gravitation, traveling at the speed of light, would take eight minutes to reach the earth after the sun disappeared. The theory of gravity and the theory of relativity were now made compatible.

Once again, many physicists greeted Einstein's new theory of gravity with skepticism. Physicists, already reeling from Einstein's statement that we live in a four-dimensional space-time continuum, were now confronted with an even more incredible theory: that this continuum is warped by the presence of matter-energy.

However, on May 29, 1919, Einstein's general theory of relativity was tested dramatically in Brazil and Africa during a total eclipse. Einstein's theory predicted that a light beam's path (like matter) should bend as it passed by the sun (see figure). This meant that the sun's vast matter-energy could somehow warp the space-time. The deflection of starlight around the sun was a dramatic verification of these ideas.

This distortion of the path of starlight was measured by comparing the position of the stars at night with their position in daytime during an eclipse, when stars become visible. World headlines were

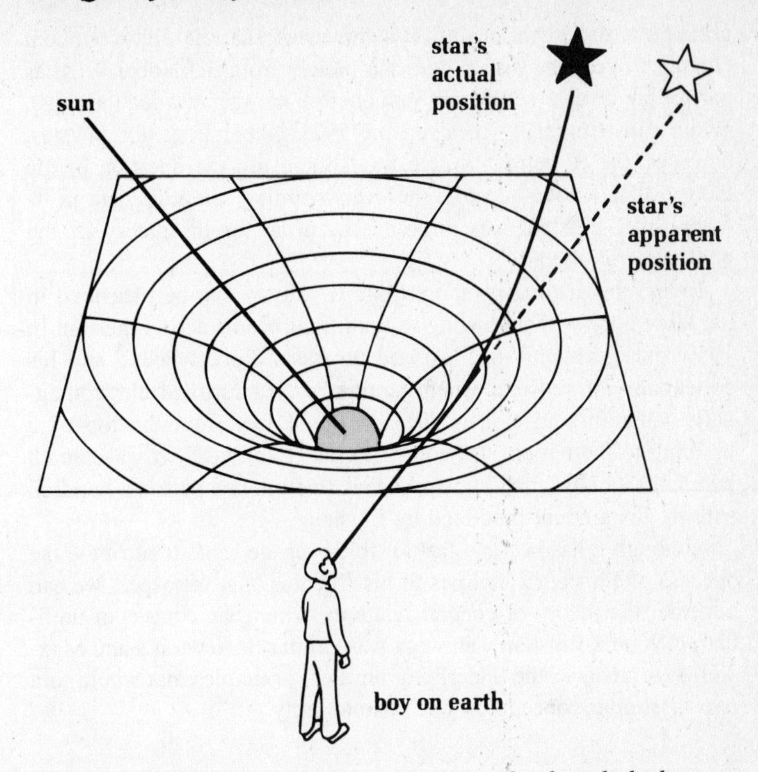

According to Einstein, gravity bends the path of starlight because the sun actually warps space-time in its vicinity. In this diagram, the black star represents the actual position of the star, whereas the white star represents its apparent position when viewed from the earth.

made when scientists measured the bending of starlight in the presence of the sun and verified the theory of general relativity.

Einstein was so sure that the physical picture and the equations were correct that he was not surprised by the results of the eclipse experiment. That year, a student asked Einstein what his response would have been if the experiment had failed. ''Then I should have been sorry for the dear Lord,'' Einstein replied, ''but the theory is correct.''[4]

(In fact, Einstein based his theories on such rigorous physical principles and such beautiful symmetries that he felt confident enough to promise his ex-wife the money from his Nobel Prize as part of the divorce settlement years before he was awarded the prize. When Einstein finally received the 1921 Nobel Prize for physics, however, the Nobel committee was so split on the question of the theory of relativity—despite the overwhelming scientific data in its favor—that Einstein was awarded the prize for his theory of the photoelectric effect.)

Today, the distortion of light due to gravity can be measured in the laboratory without having to send light beams across the sun. In 1959 and again in 1965, Harvard professor Robert Pound and his colleagues showed that when gamma rays (a form of electromagnetic radiation) traveled a distance of 74 feet from the top of a building to the bottom, the force of gravity shifted their wavelength by a fantastically small but significant amount: one part in a hundred trillion, the amount predicted by Einstein.

Although it has become fashionable over the years to attribute the success of Einstein's theories to his "genius," in retrospect we can consider the theory of general relativity within the context of unification. Again, Einstein's strategy was similar to Newton's and Maxwell's: to uncover the underlying physical principles that would join two dissimilar concepts in one cosmic unity.

From Revolutionary to Relic

Buoyed by his early successes with the theories of space-time and gravitation, Einstein set out to hunt even bigger game: the unified field theory, which would unite his geometric theory of gravity with Maxwell's theory of light.

Ironically, although the world recognizes that Albert Einstein ranks alongside Isaac Newton because he dared to penetrate the secrets of our universe, many don't realize that Einstein spent the last thirty years of his life on a solitary, frustrating, and ultimately futile quest for the unified field theory. In the 1940s and 1950s, many physicists claimed that Einstein was over the hill. They said he was isolated, out of touch, and ignorant of the new developments in atomic physics, namely the quantum theory. A few even sneered behind his back that he had become senile, a crackpot engaged in a preposterous chase after the Holy Grail. Even J. Robert Oppenheimer, director of the Institute for Advanced Studies, where Einstein worked, said to his colleagues on a number of occasions that Einstein's quest was futile.

Einstein himself confessed, "I am generally regarded as a sort of petrified object, rendered blind and deaf by the years."[5] In the last years of his life, he was almost totally isolated from his fellow physicists because he was totally absorbed by the unified field theory rather than the new developments in atomic physics and the quantum theory. "I must seem," he remarked in 1954, "like an ostrich who forever buries its head in the relativistic sand in order not to face the evil quanta."[6]

Indeed, Einstein's frustration with a few of his colleagues, some of whom he considered shortsighted and narrow, showed when he wrote, "I have little patience with scientists who take a board of wood, look for its thinnest part and drill a great number of holes where drilling is easy." He also once remarked to his secretary that physicists a hundred years later would be able to appreciate his labors, but not the current generation of physicists. (Not that the occasional loneliness bothered him too much. "The essential of the being of a man of my type," he once said, "lies precisely in *what* he thinks and *how* he thinks, not in what he does or suffers."[7])

Meanwhile, instead of trying to unite light with gravity (which most physicists thought was premature or even impossible), the world of science was riveted in an entirely new direction: the birth of atomic and nuclear physics.

In history, never has a new branch of science been heralded by such a momentous event: the detonation of the atomic bomb. Suddenly, the obscure work of a few physicists working with pencil and paper began to alter the course of human events. Their arcane equations, understood only by a handful of people working in places like the Los Alamos laboratory in New Mexico, suddenly became the pivotal force in world history.

Throughout the 1930s, 1940s, and 1950s, the dominant activity in physics was not relativity or the unified field theory but developments in the quantum theory. Most of Einstein's colleagues, such as Niels Bohr from Copenhagen and Werner Heisenberg from Göttingen, were busy constructing the mathematical language that would describe atomic and nuclear phenomena: quantum mechanics. Throughout that era, Einstein stood practically alone in his pursuit of the unification of light with gravity.

Some have argued that Einstein made the biggest blunder of his life by rejecting quantum mechanics. This, however, is a myth perpetuated by scores of science historians and journalists who are largely ignorant of Einstein's scientific thought. This myth survives only because most of these historians are not fluent in the mathematics used to describe the unified field theory.

Instead of showing how outdated he was, a careful scientific reading of Einstein's work published fifty years ago reveals that he was surprisingly modern in his approach. These papers show clearly that Einstein eventually accepted the validity of quantum mechanics. However, his personal belief was that quantum mechanics was an incomplete theory, in the same way that Newton's theory of gravity was not incorrect, merely incomplete.

Einstein believed that quantum mechanics, while highly successful, was not a final theory. His later scientific work, which has been largely ignored by nonscientists and historians, shows that he believed his unified field theory, as a by-product, would account automatically for the features of quantum mechanics. Subatomic particles and atoms, Einstein thought, would only appear as solutions to his geometric theory of gravity and light.

Einstein, however, died in the midst of his pursuit of the notion that the forces of nature ultimately must be united by some physical principle or symmetry. Even four decades after his death, most of his biographers skip over the last years of his physics research, ignoring the blind alleys he explored in his search for the unified field theory and concentrating instead on his devotion to nuclear disarmament.

Einstein's Mistake

Although physicists do not fully comprehend all the details necessary to unite the four fundamental forces into one theory, they do understand why Einstein had so much trouble wrestling with the unified field theory. We understand where Einstein went wrong.

Einstein once said that in his relativity theory he placed clocks everywhere in the universe, each beating at a different rate, but in reality he couldn't afford to buy a clock for his home. In this way, Einstein revealed a clue to the way he arrived at his great discoveries: he thought in physical pictures. The mathematics, no matter how abstract or complex, always came later, mainly as a tool by which to translate these physical pictures into a precise language. The pictures, he was convinced, were so simple and elegant that they could be understood by the general public. The mathematics might be obscure and complex, but the physical picture always should be elemental.

One of Einstein's biographers noted, "Einstein always began with the simplest possible ideas and then, by describing how he saw the problem, he put it into the appropriate context. This intuitive approach was almost like painting a picture. It was an experience that taught me the difference between knowledge and understanding."[8]

Because of Einstein's keen insight, he was able to see farther than others. It was Einstein's great pictorial insight that led him to propose the relativity theory. For three decades, he was a towering figure in physics because his physical pictures and conceptual ability were unerringly correct. The irony is, however, that in the last three decades of his life, Einstein failed to create the unified field theory largely because he abandoned this conceptual approach, resorting to the safety of obscure mathematics without any clear visual picture.

Of course, Einstein was aware of the fact that he lacked a guiding physical principle. He once wrote, "I believe that in order to make real progress one must again ferret out some general principle from nature."[9] No matter how hard he tried, however, he could not think of a new physical principle, so he gradually became obsessed with purely mathematical concepts, such as "twisted" geometries, which are bizarre mathematical structures devoid of physical content. He ultimately failed to create the unified field theory, which was to have been the centerpiece of his research, because he strayed from his original path.

In retrospect, we see that the superstring theory may be the physical framework that eluded Einstein for so many years. The superstring theory is very graphic, encompassing the infinite number of particles as modes of a vibrating string. If the theory lives up to its promise, then we see that, once again, the most profound physical theories can be summarized pictorially in a surprisingly simple fashion.

Einstein was correct in his pursuit of unification. He believed that an underlying symmetry was at the root of the unification of all forces. However, he used the wrong tactic, trying to unite the force of gravitation with the electromagnetic force (light) rather than with the nuclear force. It was natural that Einstein would try to unite these two forces, because they were the subject of intense investigation during his lifetime. However, he consciously chose to neglect the nuclear force, which is perhaps understandable because it was the most mysterious of the four forces at that time. He also felt uncomfortable with the theory that describes the nuclear force: quantum mechanics.

While relativity uncovers the secrets of energy, gravity, and space-time, the other theory that dominated the twentieth century, quantum mechanics, is a theory of matter. In simple terms, quantum mechanics successfully describes atomic physics by uniting the dual concepts of waves and particles. But Einstein didn't realize, as physicists do now, that the key to the unified field theory is found in the marriage of relativity and quantum mechanics.

Einstein was a master at understanding the nature of forces. But he was weak in his understanding of matter, especially nuclear matter. We shall now turn to this.

3

The Quantum Puzzle

By THE EARLY 1900s, the scientific world was thrown into turmoil by a series of new experiments that challenged three centuries of Newtonian physics. The world was witnessing the birth pangs of a new physics emerging from the ashes of the old order. Out of this chaos, however, emerged not one but two theories.

Einstein pioneered the first theory—relativity—and concentrated his efforts on understanding the nature of forces such as gravity and light. The foundation for understanding the nature of matter, however, was laid by the second theory, quantum mechanics, which governs the world of subatomic phenomena. It was created by Werner Heisenberg and his collaborators.

Two Giants of Physics

In many ways, the destinies of Einstein and Heisenberg were strangely interwoven, although they created theories that are universes apart. Both of German origin, they were revolutionary iconoclasts who challenged the established wisdom of their predecessors. They so thoroughly dominated modern physics that their discoveries would determine the course of physics for over half a century.

They also did their best work at astonishingly young ages. Einstein was twenty-six when he discovered relativity. Heisenberg was only twenty-four when he laid down much of the laws of quantum mechanics (having completed his Ph.D. at the age of twenty-one) and only thirty-two when he won the Nobel Prize.

Both were also steeped in the intellectual tradition that produced the flowering of the arts and sciences in Germany around the turn of the century. Most aspiring scientists who dreamed of doing first-rank physics made the mandatory pilgrimage to Germany. (One American physicist in the late 1920s, disappointed at the relatively primitive level of physics in the United States, journeyed to Göttingen, Germany, to study with the masters of quantum mechanics. This physicist, J. Robert Oppenheimer, would later go on to build the first atomic bomb.)

The destinies of these two figures also were touched by the darker side of German history, the Prussian tradition of militarism and dictatorship. When it became clear in 1933 that the fascists were beginning a period of unprecedented repression, Einstein, being Jewish, fled Nazi Germany for his life. Heisenberg, however, remained in Germany and even worked on Hitler's atomic bomb project. Indeed, the presence in Germany of world-renowned physicists such as Heisenberg helped to persuade Einstein to write his celebrated letter to President Franklin Roosevelt in 1939, urging him to build the atomic bomb. A few years ago a former agent of the OSS (forerunner to the CIA) revealed that the Allies so feared Heisenberg that they drafted elaborate plans to assassinate him if necessary to prevent the Germans from building the atomic bomb.

Not only were the personal destinies of these two men linked, but their scientific creations were also intricately related. Einstein's masterpiece was general relativity, which begins to answer questions such as: Is there a beginning and end to time? Where is the farthest point in the universe? What lies beyond the farthest point? What happened at the creation?

By contrast, Heisenberg and his colleagues, such as Erwin Schrödinger and the Danish physicist Niels Bohr, asked precisely the opposite questions: What is the smallest object in the universe? Can matter be divided into smaller and smaller pieces without limit? In posing these questions, Heisenberg and his colleagues created quantum mechanics.

In many ways, these two theories appear to be opposites. General relativity concerns the cosmic motions of galaxies and the universe, while quantum mechanics probes the subatomic world. Relativity is primarily a theory of force fields that continuously fill up all space. (The force field of gravity, for example, can be compared to gossamerlike tendrils that extend to the outer reaches of space.) Quantum mechanics, by contrast, is primarily a theory of atomic matter, which travels much slower than the speed of light. In the world of quantum mechanics, a force field only appears to fill up all space smoothly and continuously. If we could examine it closely, we would find that it actually is quantized into discrete units. Light, for example, consists of tiny packets of energy called quanta or photons.

Neither theory by itself provides a satisfactory description of nature. The fact that Einstein fruitlessly pushed his relativity theory to the breaking point shows that relativity alone cannot form a basis for the unified field theory. Nor is quantum mechanics satisfactory without relativity; quantum mechanics can be used to calculate only the behavior of atoms, not the large-scale behavior of galaxies and the expanding universe.

Merging the two theories, however, has consumed the herculean efforts of scores of theoretical physicists for the past half century. Only in the last few years have physicists finally formulated, with the help of the superstring theory, a possible synthesis of both theories.

Planck—The Reluctant Revolutionary

The quantum theory was born in 1900, when physicists found themselves mystified by something called "black body radiation." They were unable to explain, for example, why a bar of steel, when heated to high temperatures, would glow and become red hot and then white hot, or why lava is red hot when it spews from an erupting volcano.

Assuming that light was purely wavelike and could vibrate at any frequency, they found that their theories could not predict the red-hot and white-hot colors. In fact, their predictions said that the radiation should have infinite energy at high frequencies, which is impossible. This quandary was called the "ultraviolet catastrophe" (where ultraviolet simply meant high-frequency radiation), and it had puzzled scientists for years.

In 1900, the German physicist Max Planck found a solution to this problem. He was a professor in Berlin, where some of the most precise experiments on black body radiation were being performed. One Sunday he and his wife were entertaining some experimental physicists. One of them, Heinrich Rubens, casually informed Planck of his latest findings on black body radiation. After Rubens left, Planck realized that he could, via a clever mathematical trick, derive an equation that fitted Rubens's data correctly. Excited by his new theory, that evening he sent a postcard to Rubens telling him of his discovery.

When Planck presented his results to the Berlin Physical Society that month, he was extremely modest, only half believing the full implications of his own theory. He proposed that radiation was not entirely wavelike, as physicists thought, but that energy transfer occurs in definite discrete packets. In his paper in December 1900, Planck cautioned, "Experience will prove whether this hypothesis is realized in nature."[1]

Planck realized that physicists had never seen the granular nature of energy because the "size" of each packet was incredibly tiny (determined by the number $h = 6.5 \times 10^{-27}$ erg sec, now called "Planck's constant"). This number is so astronomically small that we never see quantum effects in everyday life.[2]

The physics community reacted with intense skepticism to Planck's new idea and its logical conclusion, that light was not continuous but granular. The idea that light could be chopped into "quanta" that act like a particle was considered preposterous.

Five years later, in 1905, Einstein (still an obscure physicist) carried the quantum theory to the next crucial step when he wrote down the theory of the photoelectric effect. Unlike Planck, who was a reluctant, almost timid revolutionary, and whose temperament was typical of a nineteenth-century physicist, Einstein struck out boldly in new directions with this theory.

Using Planck's strange theory of the quanta, Einstein asked what happens when a particle of light strikes a metal. If light was a particle obeying Planck's theory, then it should bounce the electrons out of some atoms in the metal and generate electricity. Einstein, using

Planck's constant, then calculated the energy of the ejected electrons.

It didn't take long for experimental physicists to verify Planck's and Einstein's equations. Planck won the Nobel Prize in 1918 for his quantum theory, followed in 1921 by Einstein for the photoelectric effect.

Today, we benefit from the applications of the quantum photoelectric effect. Television, for example, is made possible by this discovery. Television cameras utilize the photoelectric effect to record a picture on a metal surface. The light enters through the lens of the camera, hits the metal, and creates certain patterns of electricity, which are then converted into television waves that are beamed into home television sets. Unlike ordinary camera film, which can be exposed only once, this metal can be used repeatedly, and hence capture moving images.

Quantum Cookbooks

For thousands of years it was thought that particles and waves were distinct entities. By the turn of this century, however, this distinction collapsed. Not only did Planck and Einstein show that light (a wave) had definite particlelike characteristics, but experiments with electrons were showing that particles were exhibiting wavelike characteristics.

In 1923, a young French prince and physics graduate student, Louis de Broglie, wrote the basic relations that a "matter-wave" should obey, stating that an electron should have a definite frequency and a wavelength, just like light waves.

The decisive step, however, was taken in 1926 by the Viennese physicist Erwin Schrödinger. Excited by the relations written down by de Broglie, Schrödinger wrote the complete equation (called the Schrödinger wave equation) that these waves should obey. (A different, but equivalent, form of the theory was written by Heisenberg almost simultaneously.) With this, the old *quantum theory* of Planck, Einstein, and Bohr made the transition to the mature *quantum mechanics* of Schrödinger and Heisenberg.

Before 1926, scientists thought it was hopeless to try to predict the chemical properties of even the simplest compounds in the world. After 1926, however, physicists went from ignorance to almost complete understanding of the equations governing simple atoms. The power of quantum mechanics was so enormous that all of chemistry could, in principle, be reduced to a series of equations.

To a physicist, working with the Schrödinger wave equation is like cooking from an elaborate cookbook, for it tells exactly how much of this ingredient should be mixed, and how long it should be stirred, in order to determine the exact properties of atoms and molecules. Although the Schrödinger wave equation is difficult to solve for increasingly complicated atoms and molecules, we could, if we had a large enough computer, deduce the properties of all known chemicals from this equation. Quantum mechanics, however, is even more powerful than an ordinary cookbook because it also allows us to calculate the properties of chemicals that we have yet to see in nature.

The Transistor, Laser, and Quantum Mechanics

The consequences of quantum mechanics are all around us. Without quantum mechanics, a plethora of familiar objects, such as television, lasers, computers, and radio, would be impossible. The Schrödinger wave equation, for example, explains many previously known but puzzling facts, such as conductivity. This result eventually led to the invention of the transistor. Modern electronics and computer technology would be impossible without the transistor, which in turn is the result of a purely quantum mechanical phenomenon.

For example, in a metal, the atoms are arranged in orderly fashion in a lattice. The Schrödinger equation predicts that the outer electrons in the metal atoms are bound to the nucleus only loosely and, in fact, may roam freely throughout the entire lattice. Even the smallest electric fields can push these electrons around the lattice, which in turn creates the electric current. This is why metals conduct electricity. For rubber and plastics, however, the outer electrons are more tightly bound and there are no such free-roaming electrons that can create a current.

Quantum mechanics also explains the existence of materials called semiconductors, which can at times act like conductors and other times like insulators. Because of this, a semiconductor can be used as an amplifier to control the flow of electricity. Like a water faucet, where the flow of water is controlled by a simple twist of the wrist, the transistor controls the flow of electricity. Today, transistors control the flow of electricity in our personal computers, our radios, our televisions, and the like. For their invention of the transistor, three quantum physicists shared the Nobel Prize in 1956: John Bardeen, William Shockley, and Walter Brattain.

Quantum mechanics has spawned yet another invention—the laser—which is currently altering the way we conduct industry and commerce.

Quantum mechanics, first of all, explains why neon and fluorescent lamps work. In a neon light, an electric current surges through a tube of gas, energizing the atoms of the gas and kicking their electrons upstairs into a higher orbit, or energy level. The electrons in the gas atoms, which are now in an "excited" state, would like to decay back to their original state of lower energy. When the electrons finally do decay back into a lower orbit, they release energy and emit light.

In a light bulb, the excited atoms decay randomly. In fact, all the light around us, including sunlight, is random, or incoherent, radiation, a mad jumble of radiation vibrating at different frequencies and different phases. However, physicists such as Charles Townes of the University of California at Berkeley used quantum mechanics to predict that, in certain cases, the excited atoms could be made to decay *at once in precise synchronization.* This new type of radiation, called "coherent radiation," had never been seen before in nature. In 1954, Townes and his colleagues successfully produced a pulse of coherent radiation, the purest form of radiation ever seen.

Although Townes's pioneering work was for microwave radiation (for which he won the Nobel Prize in 1964), scientists quickly realized that his theories would work for light as well. Although Buck Rogers–type ray guns and beams that can blast incoming nuclear missiles are well beyond our present capability, commercial lasers can be used to cut metals, to transmit communications, and to perform surgery, and newer applications are being discovered daily. Doctors, for example, are using minilasers to send light energy along fine glass wires to burn out fatty deposits in the veins of people likely to have heart attacks. Laser discs have changed the way stereo recorders are built, and many supermarkets now use laser light at checkout counters to read instantly the black lines (bar codes) seen on the packaging of most products.

Perhaps the most spectacular commercial application of the laser would be the creation of three-dimensional television. Already, Visa cards are being issued with the "hologram" image of a three-dimensional bird. It is conceivable that in the future our television screens, instead of being flat planes, will be three-dimensional spheres in which we can see three-dimensional people moving about. Our children and grandchildren probably will watch three-dimensional television in their living rooms, compliments of quantum mechanics.

In addition to the transistor and the laser, hundreds of other important discoveries owe their existence to quantum mechanical effects. To name just a few:

- *Electron microscopes.* These exploit the wavelike properties of electrons to see objects to the size of viruses. Millions of people have benefited directly from the enormous medical applications of this quantum mechanical invention.
- *Unlocking the DNA molecule.* X-ray diffraction and other probes are used to determine the structures of these complex organic molecules. Eventually, the secret of life itself may emerge from a quantum mechanical study of these molecules.
- *Fusion machines.* These will use the nuclear reactions of the sun to create enormous amounts of energy on the earth. Although there are still many practical unsolved problems with respect to fusion machines, eventually they may provide a virtually unlimited source of power.

Without question, the success of quantum mechanics has altered the foundation of medicine, industry, and commerce. The irony is that quantum mechanics, which seems so definitive and clear-cut in its practical applications, actually is based on uncertainties, probabilities, and philosophically bizarre ideas. In short, quantum mechanics dropped a bomb on the world of physics, and the effects were shattering. "Anyone who is not shocked by quantum theory," proclaimed Niels Bohr, "has not understood it."

Heisenberg's Uncertainty Principle

In 1927, Werner Heisenberg proposed that it is impossible to know the velocity and the location of an object simultaneously. A wave, after all, is a smeared object. If we are standing on the beach, how can we calculate precisely the velocity and location of an ocean wave? We can't. One can never know precisely the position and the velocity of an electron at the same time. This is also a direct consequence of the Schrödinger equation.

According to Heisenberg, this uncertainty arises because, in the subatomic realm, the very act of observing an object changes its position and velocity. In other words, the process of taking a measurement of an atom's system disturbs the system so greatly that it alters its state, making the system qualitatively different from its state before the measurement was taken. For example, an electron is so small that to measure its position in an atom, photons of light must be bounced off of it. However, the light is so powerful that it bumps the electron out of the atom, changing the electron's position and location.

However, one would argue, with a better measuring apparatus, couldn't the velocity and the position of the electron be measured without altering it? According to Heisenberg, the answer is no. Quantum mechanics asserts that we can never know simultaneously, no matter how sensitive our measuring devices are, the exact position and velocity of a single electron. We can know one condition or the other, but not both at the same time. This is called Heisenberg's Uncertainty Principle.

Downfall of Determinism

Newton thought that the universe was like a vast cosmic clock that God wound up at the beginning of time. It's been ticking ever since, according to the three laws of motion formulated by Newton. This theory, called Newtonian determinism, states that the three laws of motion can mathematically determine the precise motion of all particles in the universe.

The French mathematician Pierre Simon Laplace took this one

step further and believed that all future events (not just the return of Halley's comet and future eclipses of the sun, but even future wars and irrational human decisions) could be calculated in advance if the initial motion of all of the atoms at the beginning of time were known. For example, determinism in its most extreme form states that it is possible to calculate in advance with mathematical precision which restaurant you will be eating in ten years from now and what you will order.

Moreover, according to this view, whether we wind up in heaven or hell is determined ahead of time. There is no free will. (When Laplace wrote his magnum opus, *Celestial Mechanics,* Napoleon is supposed to have asked him why he did not mention the Creator. Laplace answered, "I have no need for that hypothesis.")

According to Heisenberg, however, all of this is nonsense. Our fate is not sealed in a quantum heaven or hell. The Uncertainty Principle makes it impossible to predict the precise behavior of individual atoms, let alone the universe. Moreover, according to the theory, in the subatomic realm, *only probabilities can be calculated.* Since, for example, it is impossible to know the exact position and velocity of an electron, it is impossible to predict much about the electron's individual behavior. We can, however, predict with amazing accuracy the probability that a large quantity of electrons will behave a certain way.

For example, imagine the millions of students who take the college entrance exams every year. It is difficult to predict how each individual will perform on the exam, but we can determine, with uncanny accuracy, the average performance of the entire class. The bell-shaped curve, in fact, changes very little year to year. Hence, we can predict how several million students will score on the exam before they have taken it but can predict nothing about the results of any one student.

Similarly, in the case of a single radioactive uranium nucleus, which is unstable and eventually will disintegrate, it can never be predicted precisely when and with what energy it will decay. Without actually measuring the state of the nucleus, quantum mechanics cannot tell whether it is still intact or has decayed. In fact, the only way quantum mechanics can describe a single nucleus is to assume that it is a mixture of these two states. A single uranium nucleus before it is measured, then, is—as far as physicists are concerned—in a nether state between being intact and decayed. With this strange assumption, quantum mechanics allows us to calculate, with astonishingly high precision, the rate at which billions of uranium atoms will decay.

Curiosity Killed the Cat

Although scientists have never seen a violation of quantum mechanics in the laboratory (but have seen plenty of confirmations), the theory continually violates "common sense." The notions introduced by quantum mechanics are so novel that Erwin Schrödinger devised a clever "thought experiment" in 1935 that captured its apparent absurdity.

Imagine a bottle of poison gas and a cat trapped in a box, which we are not allowed to open. Obviously, although we cannot peer into the box, we can say that the cat is either dead or alive. Now imagine that the bottle of poison gas is connected to a Geiger counter, which can detect radiation from a piece of uranium ore. If a single uranium nucleus disintegrates, it releases radiation, which sets off the Geiger counter, which in turn breaks the bottle and kills the cat.

According to quantum mechanics, we cannot predict with certainty when a single uranium nucleus will disintegrate. We only can calculate the probability of billions upon billions of nuclei disintegrating. Therefore, to describe a single uranium nucleus, quantum mechanics assumes that it is a mixture of two states—one where the uranium nucleus is inert, the other where it has decayed. The cat is described by a wave function that contains the possibilities that it is both dead and alive. In other words, we must assume statistically that the cat is a mixture of two states.

Of course, once we are allowed to open the box and take a measurement, we can determine with certainty whether the cat is dead or alive. But before we open the box, according to probabilities, the cat is statistically in the nether state of being dead *and* alive. If that isn't weird enough, the very act of opening the box decides whether the cat is dead or alive. According to quantum mechanics, the measurement process itself determines the state of the cat. To make matters worse, quantum mechanics also implies that objects do not exist in a definite state (e.g., dead or alive) until they are observed.

Einstein was disturbed by the implications of quantum paradoxes such as Schrödinger's cat. "No reasonable definition of reality," he wrote, "could be expected to permit this."[3] He, like Newton before him, believed in an objective reality, holding that the physical universe exists in a precise state independent of any measuring process. The introduction of quantum mechanics opened a hornet's nest of philosophical ideas that have been buzzing around ever since.

Philosophy and Science

Scientists always have been interested in philosophy. "Science without epistemology," Einstein wrote in his later years, "is . . . primitive and muddled."[4] Indeed, as a young man, Einstein and several friends founded the Olympian Academy, an informal group organized to study philosophy. Erwin Schrödinger, a few years before he published his wave equation, decided temporarily to abandon a physics career in favor of philosophy. Max Planck wrote about free will and determinism in his book *The Philosophy of Physics.*

Although quantum mechanics has triumphed decisively in every experiment performed by scientists on the subatomic level, it raises the old philosophical question: When a tree falls in the forest, does it make any sound if there is no one to hear it? Eighteenth-century philosophers such as Bishop Berkeley and the solipsists would answer "no." To the solipsists, life was a dream, which had no material existence apart from the dreamer. A table exists only if a consciousness is there to observe it. Descartes' phrase, "I think, therefore I am," would apply to the solipsists.

On the other hand, all the great advances in science since the time of Galileo and Newton have assumed that the answer to the falling tree question is "yes"—that the laws of physics exist objectively, apart from human affairs, not subjectively, within the realm of observation.

However, the quantum physicists—basing their statements on mathematical formulas that are valid and resoundingly successful—take a philosophical leap and state that reality does not exist without a measurement taking place. In other words, the observation process creates the reality. (It should be stressed, however, that the original quantum physicists applied this philosophy only to the subatomic realm; they weren't solipsists.)

At first, traditional physicists were skeptical of this new worldview. Indeed, the founders of quantum mechanics expressed their concern, because it forced them to abandon the classical world of Newtonian physics. Heisenberg would remember conversations with Bohr late into the night in 1927 that would end "almost in despair," followed by a walk alone in the park, during which Heisenberg would repeat to himself the question: Could nature possibly be as absurd as it seemed in these atomic experiments? But eventually the quantum physicists embraced this new theory wholeheartedly, as do many physicists today, and it dominated the course of physics for the next forty-five years.

There was one physicist, however, who never accepted the quantum view of reality: Albert Einstein. He objected to quantum mechanics for several reasons. First, he did not see probabilities as a valid foundation for an entire theory. He couldn't accept the pure-chance aspect built into a theory of probabilities. "Quantum mechanics is very impressive," he wrote to Max Born, ". . . but I am convinced that God does not play dice."[5]

Second, Einstein believed that the quantum theory was incomplete. "The following requirement for a complete theory," he argued, "seems to be a necessary one: *every element of the physical reality must have a counterpart in the physical theory*"[6] (italics original). Quantum mechanics fails in this regard; dealing only with group behavior, it is a theoretical system that cannot account in detail for individual happenings.

Moreover, Einstein, a firm believer in causality, could not accept a nonobjective view of the universe. In response to the experimental success of quantum mechanics, Einstein wrote to Born: "I am convinced of [objective reality] although, up to now *success* is against it."[7] Indeed, he might have been thinking of himself when he wrote about Benedict de Spinoza: ". . . the spiritual situation with which Spinoza had to cope peculiarly resembles our own . . . he was utterly convinced of the causal dependence of all phenomena, at a time when the success accompanying the efforts to achieve a knowledge of the causal relationship of natural phenomena was still quite modest."[8]

Einstein stood almost alone in his objections. While other physicists raced to join the quantum bandwagon, he maintained until his death that the theory was incomplete. "I have become an obstinate heretic in the eyes of my colleagues,"[9] Einstein wrote to a friend. It didn't seem to disturb him much, though. "Momentary success," he observed scathingly in 1948, "carries more power of conviction for most people than reflections on principle."[10] He also wasn't swayed by majority opinion; speaking of Newton's old gravitation theory, Einstein would point out that the theory was successful for more than two centuries before it was revealed to be incomplete.

It should be emphasized that Einstein did accept the mathematical equations of quantum mechanics. However, he felt that quantum mechanics was an incomplete manifestation of an underlying theory (the unified field theory) where an objectively real description is possible. He never abandoned his search for a theory that would merge quantum phenomena with relativity. Of course, he would never live to see the day when superstrings perhaps became that theory.

Pragmatism Rules

Meanwhile, during the 1930s and 1940s, quantum mechanics was the rage, with perhaps 99 percent of the world's physicists in one camp, and Einstein staunchly holding his ground in the other.

A small minority of scientists, including Nobel Prize–winning physicist Eugene Wigner, took the position that measurement implies some sort of consciousness. Only a conscious person or entity, they argued, can perform a measurement. Therefore, according to this minority, since (according to quantum mechanics) the existence of all matter depends on measurement, the existence of the universe depends on consciousness.

This does not have to be human consciousness—it can be intelligent life somewhere else in the universe, perhaps some sort of alien consciousness, or even, as some have argued, God. Since quantum mechanics blurs the distinction between the measured and the observer, then perhaps, according to their view, the world sprang into being when the observer (a conscious being) took the first measurement.

The vast majority of physicists, however, take the pragmatic view that measurement can indeed take place without consciousness. A camera, for example, can make a measurement without being "conscious." A photon speeding across the galaxy is in an indeterminate state, but as soon as it hits a camera lens and exposes a piece of film, its state is determined. The camera eye, therefore, functions as the measurer. Before the light beam hit the camera, it consisted of a mixture of states, but exposing the film in the camera determined the precise state of the photon. Measurements can take place without the presence of a conscious observer. Observation does not imply consciousness.[11]

(By the way, the superstring theory provides perhaps the most comprehensive way of looking at Schrödinger's cat. Usually, in quantum mechanics, physicists write the Schrödinger wave function of a certain particle. However, the complete quantum mechanical description of the superstring theory requires that we write the Schrödinger wave function of the *entire universe.* Whereas previously physicists were writing the Schrödinger wave function of, say, a point particle, the superstring theory demands that we write the wave function of space-time—that is, the universe—as well as all the particles in it. This does not resolve all the philosophical problems associated with Schrödinger's cat; it merely means that the original formulation of the problem (treating the cat in a box) may be incomplete. The final resolution of the Schrödinger's cat problem may require a much more detailed understanding of the universe.)

Most working physicists, who have enjoyed fifty years of spectacular success with quantum mechanics, simply coexist with its strange philosophical implications. We are reminded of the young physicist working at Los Alamos after World War II, who asked the great Hungarian mathematician John von Neumann about a difficult mathematical problem.

"Simple," von Neumann answered. "This can be solved by using the method of characteristics."

To which the young physicist replied: "I'm afraid I don't understand the method of characteristics."

"Young man," said von Neumann, "in mathematics you don't understand things, you just get used to them."[12]

Failure of Quantum Mechanics Without Relativity

Philosophical questions aside, in the 1930s and 1940s quantum mechanics was like an unstoppable Mack truck barreling down a highway, flattening all the problems that had puzzled physicists for centuries. One brash young quantum physicist, Paul Dirac, ruffled the feathers of many chemists when he had the nerve to say that quantum mechanics could reduce *all* of chemistry to a set of mathematical equations.

However, as successful as quantum mechanics was in explaining the properties of the chemical elements, by itself it was not a complete theory. We should be careful to point out that quantum mechanics worked only when physicists used it to analyze velocities much lower than the speed of light. When attempts were made to include special relativity, this Mack truck hit a brick wall.

In this sense, the spectacular success of quantum mechanics in the 1930s and 1940s was a fluke. Electrons in the hydrogen atom typically travel at speeds one hundred times less than the speed of light. If nature had created atoms where the electron traveled at velocities near the speed of light, special relativity would become important and quantum mechanics would have been much less successful.

On the earth, we rarely see phenomena approaching the speed of light, so quantum mechanics is valuable in explaining everyday phenomena such as lasers and transistors. When we analyze the properties of ultrafast and high-energy particles in the cosmos, however, quantum mechanics can no longer ignore relativity.

Imagine, for a moment, driving a Toyota on a race track. As long as you drive the car slower than, say, 100 miles per hour, it will perform well. However, when you try to speed past 150 miles per hour, the car might break down and spin out of control. This doesn't mean that our understanding of car engineering is obsolete and must be thrown away; rather, for speeds beyond 150 miles per hour, we simply need a drastically modified car that can handle such high velocities.

Similarly, when dealing with velocities much lower than the speed of light (where special relativity can be ignored), scientists have found no deviations from the predictions of quantum mechanics. At high velocities, however, the theory fails. Quantum mechanics must be married to relativity.

The first marriage of quantum mechanics and relativity was a disaster, creating a crazy theory (called "quantum field theory") that for decades produced only a series of meaningless results. Every time physicists tried to calculate, for example, what happens when electrons collide, quantum field theory would predict infinite values for the collision.

The complete union of quantum mechanics and relativity—both special and general—has been one of the great scientific problems of this century, which only the superstring theorists claim to have solved.

Quantum mechanics alone is limited because, like nineteenth-century physics, it is still based on point particles, not superstrings.

In high school we learn that force fields such as gravity and the electric field obey the "inverse square law"—that is, the farther one distances oneself from a particle, the weaker the field becomes. The farther one travels from the sun, for example, the weaker its gravitational pull will be. This means, however, that as one approaches the particle, the force rises dramatically. In fact, at its surface the force field of a point particle must be the inverse of zero squared, which is 1/0. Expressions such as 1/0, however, are infinite and ill-defined. The price we pay for introducing point particles into our theory is that all our calculations of physical quantities, such as energy, are riddled with 1/0s. This is enough to render a theory useless; calculations with a theory plagued with infinities cannot be made because the results cannot be trusted.

The problem of infinities would haunt physicists for the next half century. Only with the advent of the superstring theory has this problem been solved, because superstrings banish point particles and replace them with a string. The original assumption made by Heisenberg and Schrödinger—that quantum mechanics should be based on point particles—was simply too stringent. A new quantum mechanics can be built on a theory of superstrings.

The mechanism by which the theory manages to marry both special and general relativity with quantum mechanics, however, is a fascinating feature found only in strings, which we shall discuss in the following chapters.

4

The Riddle of Infinities

WHAT DO safecrackers and theoretical physicists have in common? Richard Feynman was an accomplished safecracker who broke into some of the most closely guarded safes in the world. He also was a world-renowned physicist. According to Feynman, both the safecracker and the physicist are adept at sifting through seemingly random clues and piecing together subtle patterns that hold the answers to the problem.

Since the 1930s, physicists have been consumed by the frustrating task of cracking the "safe" of quantum field theory and finding the key to the successful marriage between quantum mechanics and relativity. Only in the past twenty years, however, have physicists realized that the tantalizing clues found in the experimental data from atom smashers form a systematic pattern.

Today, we realize that this pattern can be expressed as an underlying mathematical symmetry that links forces that appear to be totally dissimilar. These symmetries, we shall see, will play a central role in canceling the divergences found in quantum field theory. The discovery that these symmetries can cancel these divergences is perhaps the greatest lesson of the past half century in physics.

Feynman the Prankster

This knack of exploiting symmetries and isolating the key factors in any problem led Feynman to produce the first successful union of special relativity with quantum mechanics in 1949, for which he and his colleagues won the Nobel Prize in 1965.

The theory, called "quantum electrodynamics (QED)," was a modest contribution by today's standards, treating only the interactions of the photon (light) and the electron (and not the weak or nuclear force and certainly not gravity); but it marked the first major advance, after years of frustration, in uniting special relativity with quantum mechanics.

The QED theory was as different from relativity as Feynman's personality was different from Einstein's. Unlike most other physicists, Einstein had a playful streak and took every opportunity to poke fun at the stuffy totems of conventional society. But if Einstein was playful, physicist Richard Feynman was an outlandish prankster.

Feynman's early fascination with practical jokes surfaced while he was a young physicist working on the atomic bomb project in the 1940s. Priding himself on his abilities as a safecracker, one day he cracked three vaults in a row at Los Alamos that contained the sensitive military equations of the atomic bomb. In one vault he left a message scribbled on yellow note paper bragging how easy it was to crack open the safe, and signed the message "Wise Guy." In the last vault, he put in a similar message and signed it "Same Guy."

The next day, Dr. Frederic de Hoffman opened the safes and found these mysterious messages lying atop the most closely guarded secrets in the world. Feynman recalled: "I had read in books that when somebody is afraid, his face gets sallow, but I had never seen it before. Well, it's absolutely true. His face turned a gray, yellow green—it was really frightening to see."[1] Dr. de Hoffman read the sheet signed by the mysterious "Same Guy" and immediately yelled, "It's the same guy who's been trying to get into Building Omega!" In his hysteria, Dr. de Hoffman falsely concluded that the safecracker was the same man who was apparently spying on another top-secret project at Los Alamos. Feynman soon confessed to being the culprit.

Feynman's renowned talent for unlocking safes apparently came in handy when he was tackling a much more difficult problem: eliminating the infinities from the quantum field theory.

The S-Matrix—Why Is the Sky Blue?

When Feynman was a student at the Massachusetts Institute of Technology, he asked himself a simple question: What is the most important problem in all of theoretical physics? Clearly, it was the elimination of the infinities infesting the quantum field theory.

Feynman set out to predict numerically what happened when particles, such as electrons or atoms, bumped into one another. When describing such collisions, physicists use the term *S-matrix (s* stands for "scattering"), which is merely a set of numbers that contains all the information of what happens when particles collide. It tells us how many particles will scatter at a certain angle with a certain amount of energy.

Calculating the S-matrix is profoundly important because if the S-matrix were known completely, in principle it would be possible to predict virtually *all* the properties of the material.

One importance of the S-matrix is that it can explain puzzling, everyday phenomena. For example, physicists in the nineteenth century, using a crude form of the S-matrix for the scattering of sunlight in the air, were able to explain for the first time why the sky was blue and sunsets were red.

When we look at the sky during the daytime, we mainly see sunlight that has bounced off air molecules and scattered in all directions in the atmosphere. Because blue light scatters more easily than red light, and because the light from the sky is mostly scattered light, the sky appears blue. (If, however, we imagined a world without air, the sky would look dark even in daylight, because there would be no scattered light. On the moon, which has no air to scatter the sunlight, the sky appears black, even in the daytime.)

The sunset looks red, however, due to the opposite effect: We are looking mainly at the sun itself and not at scattered light. At sunset, the sun sits near the horizon, so light from the setting sun must travel horizontally to reach our eyes and thus travels through a relatively large amount of air. By the time sunlight reaches us, only the reds are left unscattered.

Similarly, when quantum physicists of the 1930s calculated the S-matrix for colliding hydrogen and oxygen atoms, they could show that water would be created. In fact, if we knew the S-matrix for all possible collisions between atoms, in principle we could predict the formation of all possible molecules, including DNA molecules. Ultimately, this means that the S-matrix holds the key to the origin of life itself.

The fundamental problem facing physicists, however, was the fact that quantum mechanics became useless when extended to velocities near that of light. As early as 1930, J. Robert Oppenheimer found that quantum mechanics, when wedded to special relativity, predicted a useless series of infinities for the S-matrix. He wrote that unless these infinities could be eliminated, the theory must be discarded.

In the 1940s, Feynman, using his best safecracking techniques, doodled on scraps of paper, tracing pictorially what happened when electrons collided with one another. Since each doodle was actually a shorthand notation for a tremendous amount of tedious mathematics, Feynman was able to condense hundreds of pages of algebra, and isolate the troublesome infinities. These mathematical doodlings allowed him to see farther than those who were lost in a jungle of complex mathematics.

Not surprisingly, ''Feynman diagrams'' were a source of controversy within the physics community, which was split on how to deal with them. Because Feynman could not derive his rules, his critics thought that these diagrams were silly or perhaps just another of his famous practical jokes. Some of his critics preferred another version of QED being formulated by Julian Schwinger of Harvard University and Shinichiro Tomonaga of Tokyo. However, the more perceptive physicists realized that Feynman was on to something potentially profound with these pictures. Princeton physicist Freeman Dyson explained the source of this confusion:

> The reason Dick's physics was so hard for ordinary people to grasp was that he did not use equations. The usual way theoretical physics was done since the time of Newton was to begin by

> writing down some equations and then to work hard calculating solutions of the equations. . . . Dick just wrote down the solutions out of his head without ever writing down the equations. He had a physical picture of the way things happen, and the picture gave him the solutions directly with a minimum of calculation. It was no wonder that people who had spent their lives solving equations were baffled by him. Their minds were analytical; his was pictorial.[2]

Feynman's doodlings were important because they allowed him to exploit the full power of gauge symmetry, which started a revolution in physics that continues today.

Tinker Toys and Feynman Diagrams

Think of playing with Tinker Toys. Assume that there are only three types: a straight stick (a moving electron), a wavy stick (a moving photon), and a joint that can connect a wavy stick with two straight ones (representing the interaction).

Now assume that we connect these Tinker Toys in all possible ways. For example, start with the collision of two electrons. Very simply, we can use these Tinker Toys to create an infinite sequence of diagrams that describe how two electrons collide.

Of course, these diagrams are deceptively simple. There are an infinite number of these Feynman diagrams, each representing a definite mathematical expression, which, when added up, produces the S-matrix. But with a little practice, even a person with no physics background can create hundreds of Tinker Toy diagrams that describe how two electrons collide.

In essence, two types of Tinker Toy diagrams can be assembled: "loops," such as the last diagram in the figure, and "trees," which contain no loops but resemble tree branches, like the first diagram. Feynman found that the trees were finite and yielded experimentally good results. But the loop diagrams were troublesome, yielding meaningless infinities. They were divergent because the theory was still based on *point particles.* In essence, quantum physicists in the 1940s were rediscovering the problem identified by nineteenth-

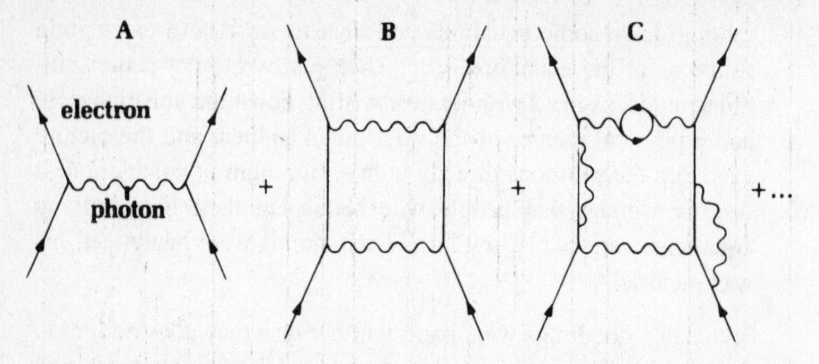

According to Feynman, when two electrons (represented by straight sticks) collide, they exchange photons (represented by wavy sticks). In diagram A, these colliding electrons exchange a single photon; in diagram B, they exchange two photons; in diagram C, they exchange many photons.

century physicists, who found that the energy of a point particle was 1/0.

Today, the superstring theory can very likely eliminate all these divergences, not just for electrons and photons but even for the gravitational interactions. However, Feynman in the 1940s found a partial solution to the infinity problem in QED.

Feynman's solution was quite novel, albeit controversial. QED is a theory that has two parameters—the charge and the mass of the electron. In addition to special relativity, Maxwell's equations possessed yet another symmetry, called "gauge symmetry,"* which allowed Feynman to regroup a large set of diagrams until he found that he could simply *redefine* the charge and mass of the electron to absorb or cancel the infinities.

At first, this juggling of infinities was greeted with intense skepticism. After all, Feynman's method assumed that the original mass

* A wave equation is defined at every point in space and time. If the equation remains unchanged when we make the same rotation at every point in space and time, then the equation has global symmetry. However, if the equation remains the same when we make a *different* rotation at *each* point in space and time, it possesses a more complex symmetry called local or gauge symmetry. We now know that gauge symmetry is probably the only way to eliminate the undesirable features of the quantum field theory.

and the charge of the electron (the "bare" mass and charge) were essentially infinite to start with, but they absorbed (that is, "renormalized") the infinities emerging from the graphs and then became finite.

Can infinity minus infinity yield a meaningful result? (Or, in the language of physics, can $\infty - \infty = 0$?)

To the critics, using one set of infinities (arising from loops) to cancel another set of infinities (coming from the electric charge and mass) looked like a parlor trick. In fact, Dirac for years criticized the renormalization theory as too clumsy to represent a genuinely profound leap in our understanding of nature. To Dirac, the renormalization theory was like a cardshark rapidly shuffling his deck of Feynman diagrams until the cards with the infinities mysteriously disappeared.

"This is just not sensible mathematics," Dirac once said. "Sensible mathematics involves neglecting a quantity when it turns out to be small—not neglecting it because it is infinitely great and you do not want it!"[3]

However, the experimental results were undeniable. In the 1950s, Feynman's new theory of renormalization (which provided a way to absorb the infinities) allowed physicists to calculate with incredible precision the energy levels of the hydrogen atom. No other theory came close to approaching the fantastic calculational accuracy of QED. Although the theory works only for electrons and photons (and not for weak, strong, or gravitational forces), it was undeniably a stunning success.

After it was demonstrated that Feynman's version was equivalent to Schwinger's and Tomonaga's, the three shared the Nobel Prize in 1965 for eliminating the infinities from QED. In hindsight, we realize that the real accomplishment was their exploiting Maxwell's gauge symmetry, which is crucially responsible for the seemingly miraculous cancellations of infinities in QED. This interplay between symmetry and renormalization, found over and over again, is one of the great mysteries of physics. These powerful symmetries are the reason why the superstring, which has the largest set of symmetries ever found in physics, has such wondrous properties.

Failure of Renormalization Theory

In the 1950s and 1960s, Feynman's rules were the rage of physics. The blackboards of the nation's top laboratories, once filled with dense equations, now blossomed with pictures filled with trees and loops. It seemed that everyone suddenly became an expert in doodling on scraps of paper and constructing Tinker Toy–like diagrams. Physicists reasoned that if Feynman's rules and renormalization theory were so successful in solving QED, then maybe lightning would strike twice and the strong and weak forces also could be "renormalized."

The 1950s and 1960s, however, were confusing decades marked by false starts. Feynman's rules were not enough to renormalize the strong and weak interactions. Physicists, not realizing the importance of gauge symmetry, explored hundreds of blind alleys without success.

Finally, after two decades of chaos, the key breakthrough was made in the weak interactions. For the first time in almost one hundred years, since the time of Maxwell, the forces of nature took another step toward unification. Once again, the key to the puzzle would be gauge symmetries.

Renormalization and Weak Interactions

Weak interactions concern the behavior of electrons and their partners, called "neutrinos." (Weakly interacting particles are collectively called "leptons.") Of all the particles in the universe, the neutrino is perhaps the most curious, because it is by far the most elusive. It has no charge, probably has no mass, and is exceedingly hard to detect.

It was predicted by Wolfgang Pauli on purely theoretical grounds in 1930 to explain the strange loss of energy found in radioactive decay. Pauli conjectured that the missing energy was carried off by a new particle that could not be seen in the experiments.

In 1933, the great Italian physicist Enrico Fermi published the first comprehensive theory of this elusive particle, calling it the "neutrino" ("little neutral one" in Italian). However, because the entire idea of the neutrino was so speculative, his paper originally was rejected for publication by the British journal *Nature*.

Neutrino experiments were notoriously difficult because neutrinos are very penetrating and leave no traces of their presence. In fact, they can easily penetrate through the earth. Every second our bodies are riddled with neutrinos that entered the earth through China, penetrated the earth's core, and came up through the floor. In fact, if our entire solar system were filled with solid lead, some neutrinos would be able to penetrate even that formidable barrier.

The neutrino's existence was finally confirmed in 1953 in a difficult experiment that involved studying the enormous radiation created by a nuclear reactor. Since its discovery, over the years inventors have tried to think of practical uses for the neutrino. The most ambitious would be the building of a neutrino telescope.

With the telescope we could probe directly into hundreds of miles of solid rock, which would enable us to discover new deposits of oil and rare minerals. By penetrating the earth's crust and mantle, we would be able to discover the origin of earthquakes and possibly to predict them. The idea of a neutrino telescope is a good one, but there is one catch: Where do we find the photographic film that can stop neutrinos? Any particle that can penetrate trillions of tons of rock could just as easily penetrate photographic film.

(Another suggestion would be to create a neutrino bomb. Physicist Heinz Pagels writes that it is "a pacifist's favorite weapon. Such a bomb, which could easily be as expensive as a conventional nuclear weapon, would explode with a whimper and flood the target area with a high flux of neutrinos. After terrifying everyone, the neutrinos would fly harmlessly through everything."[4])

In addition to the neutrino, the mystery of the weak interactions deepened with the discovery of other weakly interacting particles, such as the "muon." Back in 1937, when this particle was discovered in cosmic ray photographs, it looked just like an electron but was more than two hundred times heavier. For all intents and purposes, it was just a heavy electron. Physicists were disturbed that the electron seemed to have a useless twin, except heavier. Why did nature create a carbon copy of the electron? Wasn't one enough? Columbia physicist and Nobel laureate Isidor Isaac Rabi, when told of the discovery of this redundant particle, exclaimed, "Who ordered that?"

To make matters worse, physicists in 1962, using the atom smasher in Brookhaven, Long Island, showed that the muon, too, had its own distinct partner, the muon neutrino. In 1977–78 experiments at Stanford University and in Hamburg, Germany, showed that there was yet another redundant electron, this time weighing in at thirty-five hundred times the electron mass. It was dubbed the "tau" particle, with its own separate partner, the tau neutrino. Now there were three types of electrons, each with its own neutrino, each identical to the electron family except for mass. Physicists' faith in the simplicity of nature was shaken by the existence of three redundant pairs, or "families," of leptons.

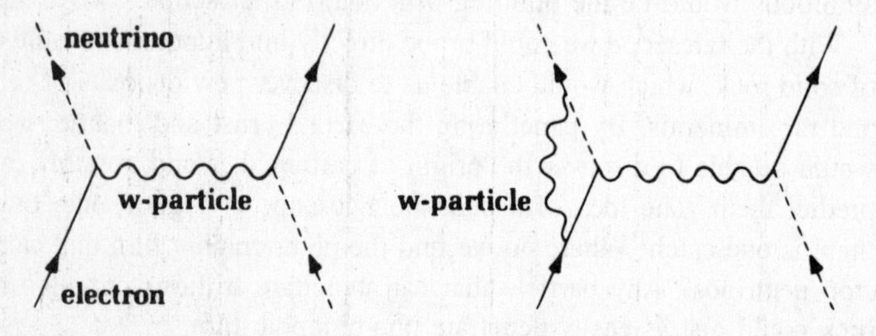

According to the W-particle theory, an electron (represented by a straight stick) collides with a neutrino (dotted stick) and exchanges a series of W-particles (wavy lines).

Faced with the problem of weak interactions, physicists used a time-honored technique: applying analogies stolen from previous theories to create new theories. The essence of QED explained the force between electrons as the exchange of photons. By the same reasoning, physicists conjectured that the force between electrons and neutrinos was caused by the exchange of a new set of particles, called W-particles (*w* for "weak").

The resulting theory (with electrons, neutrinos, and W-particles) can be explained with three kinds of Tinker Toys: a straight stick (representing the electron), a dotted stick (representing the neutrino), and a spiral (representing the W-particle), as well as a joint interaction. When electrons collide with neutrinos, they simply exchange a W-particle.

Again, after a little practice, it is not hard to assemble hundreds of Feynman diagrams for weak-interaction processes created by the exchange of the W-particle.

The problem, however, was that the theory was *nonrenormalizable.* No matter how cleverly Feynman's bag of tricks was used, the theory still was plagued with infinities. The problem was that the W-particle theory itself had a fundamental sickness—it had no gauge symmetries, as in Maxwell's equations.

As a consequence, the theory of weak interactions languished for three decades. Not only were experiments difficult to perform (because of the notoriously elusive neutrino), but the W-particle theory was unacceptable. Physicists puttered with the theory over the decades, but no significant breakthroughs were made.

Success of the Electro-Weak Theory

In 1967–68, Steven Weinberg, Abdus Salam, and Sheldon Glashow noticed the amazing similarity between the photon and the W-particle. Then they made the following observation: Although Einstein had tried to unite light with the gravitational force, perhaps the correct unification scheme was to unite the photon with the W-particle of weak interactions. This new W-particle theory, called the electro-weak theory, differed decisively from previous W-particle theories because it used the most sophisticated form of gauge symmetry available at that time, the Yang-Mills theory. This theory, formulated in 1954, possessed more symmetries than Maxwell ever dreamed of. (We will explain the Yang-Mills theory in chapter 7.)

The Yang-Mills theory contained a new mathematical symmetry (represented mathematically as $SU(2) \times U(1)$) that allowed Weinberg and Salam to unite the weak and electromagnetic forces on the same footing. This theory also treated the electron and the neutrino symmetrically as one "family." As far as the theory was concerned, the electron and the neutrino were actually two sides of the same coin. (The theory did not, however, explain why there were three redundant electron families.)

Although the theory was the most ambitious and advanced theory of its time, it raised few eyebrows. Physicists assumed that it was probably nonrenormalizable, like all the other dead ends, and therefore riddled with infinities.

Weinberg, in his original paper, speculated that the Yang-Mills version of the W-particle theory was probably renormalizable, but no one believed him. However, all this changed in 1971.

After three decades of agonizing over the infinities festering within the W-particle theory, a dramatic breakthrough was made when a twenty-four-year-old Dutch graduate student, Gerard 't Hooft, proved that the Yang-Mills theory was renormalizable. To double-check his calculation showing the cancellation of infinities, 't Hooft placed the calculation on computer. One can imagine the excitement that 't Hooft must have felt while awaiting the results of his calculation. He later recalled: "The results of that test were available by July 1971; the output of the program was an uninterrupted string of zeros. Every infinity canceled exactly."[5]

Within months, hundreds of physicists rushed to learn the techniques of 't Hooft and the theory of Weinberg and Salam. For the first time, real numbers, not infinities, poured out of the theory for the S-matrix. Earlier, from 1968 to 1970, not a single paper published by a physicist referred to Weinberg and Salam's theory. By 1973, however, when the impact of their results was being appreciated, 162 papers on their theory were published.

Somehow, in ways that physicists still don't completely understand, the symmetries built in to the Yang-Mills theory eliminated the infinities that had plagued the earlier W-particle theory. Here was the stunning interplay between symmetry and renormalization (which we shall elaborate on in chapter 7). It was also a replay of the discovery made by physicists studying QED years earlier—that symmetries somehow canceled the divergences in a quantum field theory.

Glashow—The Revolutionary Anarchist

Steven Weinberg and Sheldon Glashow were classmates from the celebrated Bronx High School of Science in New York, where they were best friends and contributed articles to the Science Fiction

Club's magazine. The Bronx High School of Science has produced three Nobel Prize winners in physics—more than any other high school in the world.

Although Weinberg and Glashow were led to the same conclusions concerning unification, they have opposite temperaments. One of their friends told *The Atlantic Monthly,* ''Steve is a royalist, Shelly a revolutionary anarchist. Steve works best by himself, Shelly works best with others. He's a futzer. He arrives in the morning with four or five wild ideas, most of them wrong, and expects other people to tear them apart. Steve is sensitive and private, while Shelly is gregarious. . . .''[6]

Glashow may be a wild ''revolutionary anarchist'' in his style, but the way in which he arrives at his ideas is by constantly bubbling forth with new ones, many of them crazy and impossible, but some of them genuine breakthroughs in physics. Of course, he relies on the help of others to shoot down the bad ideas, but nonetheless he possesses that creative instinct that many lack. In theoretical physics, simply being brilliant is not enough. One must also be able to generate new ideas, some of them bizarre, which are essential to the process of scientific discovery.

Glashow also likes to invent new particles to upset the physics establishment. After he proposed one particularly unusual particle, his collaborator Howard Georgi said, ''It was another way for him to throw rocks at the establishment.''[7]

(Glashow also has a reputation as an eccentric professor. When Michio was an undergraduate at Harvard, he took a course in classical electrodynamics from Glashow. In the middle of the final exam, while all the students were sweating to complete the problems, Glashow blurted, ''Oh, by the way, I haven't been able to solve problem number five myself. If any of you ever find the answer, please tell me.'' Everyone in the class just stared at one another in amazement.)

Glashow, in his 1979 Nobel Prize acceptance speech for the electro-weak theory, summed up the tremendous excitement of seeing the unification of subatomic forces emerge before his eyes:

> In 1956, when I began doing theoretical physics, the study of elementary particles was like a patchwork quilt. Electrodynamics,

> weak interactions, and strong interactions were clearly separate disciplines, separately taught and separately studied. There was no coherent theory that described them all. Things have changed. Today we have what has been called a standard theory of elementary physics, in which strong, weak, and electromagnetic interactions all arise from a [single] principle. . . . The theory we now have is an integral work of art: the patchwork quilt has become a tapestry.[8]

Mesons and the Strong Force

Physicists, already dizzy from the monumental success of the electro-weak theory, turned their attention to solving the strong force.

Would lightning strike three times?

Gauge symmetry had canceled the divergences of QED and the electro-weak theory. Was gauge symmetry also the key to canceling the infinities of the strong interactions? The answer was yes, but only after a considerable amount of confusion that lasted for decades.[9]

The origins of strong interaction theory date back to 1935, when Japanese physicist Hideki Yukawa proposed that protons and neutrons were held together in the nucleus by a new force created by the exchange of particles called "pi mesons." Just as in QED, where the exchange of photons between the electron and the nucleus held the atom together, Yukawa by analogy proposed that the exchange of these mesons held the nucleus together. He even predicted the mass of these hypothetical particles.

Yukawa was the first to argue that the short-range forces in nature could be explained by the exchange of massive particles. In fact, Yukawa's meson idea provided the original inspiration for other physicists to propose the W-particle a few years later as the carrier of the weak force.

In 1947, English physicist Cecil Powell discovered the meson in his cosmic ray experiments. The particle had a mass very close to that predicted by Yukawa twelve years earlier. For this pioneering work in unraveling the mysteries of the strong force, Yukawa was awarded the Nobel Prize in 1949, and Powell received the prize the following year.

Although this meson theory met with considerable success (and was renormalizable as well), it was not by any means the final word. In the 1950s and 1960s, physicists using atom smashers in laboratories across the country were discovering hundreds of different types of strongly interacting particles, now called "hadrons" (which include both the mesons and other strongly interacting particles such as the proton and the neutron).

The existence of hundreds of hadrons was an embarrassment of riches. No one could explain why nature suddenly became more, not less, complicated as scientists probed the subnuclear realm. Everything seemed so simple by contrast in the 1930s, when it was thought that the universe was built from just four particles and two forces (the electron, proton, neutron, neutrino, and light and gravity). By definition, elementary particles should be few in number, but physicists in the 1950s were flooded with new hadrons discovered in the nation's laboratories. Obviously, a new theory was required to make some sense out of this chaos.

Nobel laureate Enrico Fermi, observing the plethora of new hadrons, each one with a strange-sounding Greek name, once lamented, "If I could remember the names of all these particles, I would have been a botanist."[10]

J. Robert Oppenheimer said in jest that the Nobel Prize should be given to the physicist who *didn't* discover a new particle that year.

By 1958, the number of strongly interacting particles had grown so rapidly that physicists at the University of California at Berkeley published an almanac to keep track of them. The first almanac was nineteen pages long and categorized sixteen particles. By 1960, there were so many particles that a considerably expanded almanac, including a wallet card, was published. By 1995, the list was expanded to over 2,000 pages, describing hundreds of particles.

The Yukawa theory, although renormalizable, was still too primitive to explain the particle zoo emerging from the laboratories. Apparently, renormalizability was not enough. As we saw earlier, the missing ingredient in the W-particle theory was the gauge symmetries of the Yang-Mills theory. After decades of confusion, the same lesson, exploiting the power of gauge symmetries, would apply to the strong force as well.

Worlds Within Worlds

Physicists, searching for an analogy, were reminded of the confusion chemists faced in the 1800s. Back then, chemists asked how anyone could possibly make sense of the billions of chemical compounds that were known to exist. The first breakthrough occurred in 1869, when the Russian chemist Dmitri Mendeleev showed how all these compounds could be reduced to a set of simple elements that could be arranged in a beautiful chart, called the Mendeleev periodic table. This chart, which every high school student learns in chemistry class, suddenly made order out of chaos.

Mendeleev knew of the existence of only sixty elements. (More than one hundred are known today.) He found many missing "holes" in his table, however, which allowed him to predict the existence and properties of new elements that hadn't yet been discovered. The actual discovery of these missing elements, just where Mendeleev had predicted them, was a confirmation of his periodic chart.

In the 1930s, quantum physicists showed how even the periodic table could be explained with just three particles, the electron, the proton, and the neutron, which obeyed the laws of quantum mechanics. Of course, reducing several billion compounds down to the one hundred or so elements in the periodic chart and then down to just three particles was a significant leap in our understanding of nature.

The question now being asked was: Would the same technique work with the hundreds of hadrons being discovered in our laboratories? The key would be to discover a symmetry that would make sense out of the data.

In the 1950s, the first crucial observation was made by a group of physicists in Japan, whose most vocal spokesman was Shoichi Sakata of Nagoya University. The Sakata group, citing the philosophical works of Hegel and Engels, claimed that there should be a sublayer beneath the hadrons consisting of even smaller subnuclear particles. Sakata claimed that the hadrons should consist of three of these particles and that the meson should consist of two of these particles. His group even proposed that these subparticles obeyed a new type of symmetry, called SU(3), which describes the mathematical way in which these three subnuclear particles could be shuffled. This mathematical symmetry, SU(3), allowed Sakata and his group to make precise mathematical predictions about the layer beneath the hadron.

The Sakata school argued on philosophical and mathematical grounds that matter should consist of an infinite set of sublayers. This is sometimes called the worlds within worlds or onion theory. According to dialectical materialism, each layer of physical reality is created by the interaction of poles. For example, the interaction between the stars creates the galaxies. The interaction between the planets and the sun creates the solar system. The interaction between the atoms creates the molecules. The interaction between the electron and the nucleus creates the atom. And finally, the interaction between the proton and the neutron creates the nucleus.

However, the experimental data at the time was too crude to test their predictions. Not enough was known in the 1950s about the specific properties of all these exotic particles to confirm or invalidate the theories of the Sakata school. (Moreover, although Sakata was on the right track, it turns out that he mistakenly thought that the three fundamental particles were the proton, the neutron, and a new particle called the lambda.)

The next breakthrough for the belief that a sublayer existed beneath the hadrons came in the early 1960s, when Murray Gell-Mann of the California Institute of Technology and Israeli physicist Yuval Neéman showed that these hundreds of hadrons occurred in patterns of eight, much like Mendeleev's periodic chart. Gell-Mann even whimsically called this mathematical theory the Eightfold Way, the name of the Buddhist doctrine describing the path to wisdom. (He meant the title as a "colossal joke.") By looking for "holes" in his Eightfold Way chart, Gell-Mann—like Mendeleev before him—could predict the existence and even the properties of particles that hadn't yet been discovered.

But if the Eightfold Way was comparable to the Mendeleev periodic chart, what was the counterpart of the electron and the proton, which make up the atoms in the chart?

Later, Gell-Mann and George Zweig proposed the complete theory. They discovered that the Eightfold Way arises because of the existence of subnuclear particles (which Gell-Mann dubbed "quarks" after James Joyce's *Finnegan's Wake).* These particles obeyed the symmetry SU(3), which the Sakata school had pioneered years earlier.

Gell-Mann found that by taking simple combinations of three quarks, he could miraculously explain the hundreds of particles found in the laboratories and, more important, predict the existence of new ones. (Gell-Mann's theory, although resembling Sakata's in many ways, used a slightly different set of combinations from Sakata's, thereby correcting a small but important mistake in the Sakata theory.) In fact, by properly combining these three quarks, Gell-Mann was able to describe virtually all the particles emerging in the laboratories. For his contributions to strong interaction physics, Gell-Mann received the Nobel Prize in 1969.

As successful as the quark model was, however, one nagging question remained: Where was a satisfactory renormalizable theory that could explain the force that held these quarks together? The quark theory was still incomplete.

Quantum Chromodynamics

Meanwhile, in the early 1970s, the excitement over Weinberg and Salam's electro-weak theory was spilling over into the quark model. The natural question was: Why not try symmetry and the Yang-Mills field to eliminate the divergences?

Although the results are still not conclusive, today there is practically universal consensus that the Yang-Mills theory, with all its wondrous properties and symmetry, can successfully bind the quarks into a renormalizable framework. Under certain circumstances, a Yang-Mills particle called the "gluon" can act as if it were a sticky, gluelike substance that binds the quarks. This is called the "color" force, and the resulting theory, called "quantum chromodynamics" (QCD for short), is widely believed to be the final theory of strong interactions. Preliminary computer programs indicate that the Yang-Mills field indeed binds the quarks.

With the success of the Yang-Mills theory and QCD, physicists asked: Could nature really be this simple? By now physicists were drunk with success. The magic formula, using gauge symmetry (in the form of the Yang-Mills theory) to create renormalizable theories, seemed to be a prescription for certain success.

The next question was: Would lightning strike a *fourth* time and create a unified theory of strong, weak, and electromagnetic interactions? The answer again appeared to be yes.

5

The Top Quark

IN JULY 1994, physicists held up their champagne glasses in laboratories around the world. The elusive ''top quark'' finally had been discovered. Physicists at the Fermi National Laboratory outside Chicago could hardly contain their excitement when the press releases were handed out.

Almost immediately, *The New York Times* trumpeted this discovery on the front page. Not in recent memory had the discovery of a new subatomic particle been reported on the front page of newspapers across the country. All of a sudden, millions of people who didn't have the slightest understanding of (or even interest in) atomic physics were asking the question, ''What is a top quark?''

NBC-TV news in New York asked random people around town if they knew what a top quark was. (After some hilarious guesses, one person made a surprisingly accurate on-the-spot answer.) Comedians began to work the top quark into their stand-ups. The top quark was the first particle to get its fifteen minutes of fame!

Hunting the Top Quark

What made the top quark significant was that it was the last quark necessary to complete the ''Standard Model''—the current and most successful theory of particle interactions. To a particle physicist, it was the last and crowning achievement of a half century of painstaking effort to decode the mysteries of the subatomic world. One chapter in particle physics had been closed. A new chapter in physics was dawning.

Physicists had been searching for this elusive particle since 1977, soon after the "bottom quark" was discovered at the Fermi National Laboratory. However, over the past fifteen years searches had failed to detect the presence of the even heavier top quark. Physicists were getting nervous. If the top quark did not exist, then elementary particle physics would collapse like a house of cards. At international conferences of particle physicists, it was almost becoming a joke that experiment after experiment failed to turn up the top quark.

As Nobel laureate Steve Weinberg said, "There was tremendous theoretical expectation that the top quark is there. A lot of us would have been embarrassed if it were not."[1]

To snare the top quark, the particle accelerator at the Fermi lab, called the Tevatron, created two highly energetic beams of subatomic particles whipping around a large, circular tube, but traveling in opposite directions. The first beam consisted of ordinary protons. The other beam, circulating in the opposite direction and below the first beam, consisted of antiprotons (the antimatter twin of the proton that carries a negative electrical charge). The Tevatron then merged these two circulating beams, smashing the protons into the antiprotons at energies of almost 2 trillion electron volts. The colossal energy released by this sudden collision released a torrent of subatomic debris.

Using a battery of complex automatic cameras and computers, physicists then analyzed the debris from over a trillion photographs. To the unaided eye, these pictures look like a spiderweb, with long, curved fibers emanating from a single point. To the trained eye, these fibers represented the tracks of subatomic particles blasted out from the collision. Teams of physicists then labored over the data, sifting the photographs until just twelve collisions were selected that had the "fingerprint" of a top quark collision.

The physicists then estimated that the top quark had a mass of 174 billion electron volts, making it the heaviest elementary particle ever discovered. In fact, it is so heavy that it is nearly as massive as a gold atom (which contains 197 neutrons and protons). By contrast, the bottom quark has a mass of 5 billion electron volts.

Given the enormous stakes, and the mountains of data required to verify its existence, the physicists at the Fermi lab were careful to say that their evidence for the top quark was not conclusive. Indeed, the top quark was so massive and elusive that the combined heroic efforts of over 440 scientists from thirty-six institutions were needed to snag it. (This prompted jokes about how many physicists were needed to screw in a quark light bulb.) Even then, they hedged their bets by saying that there was a one-in-400 chance that they were wrong.

"We're in that middle ground where the excess of events that we see is too large to ignore, but too small to cry Eureka,"[2] admitted William Carithers, one of the spokesmen for the group.

Finally, eight months later, the group (and also a rival group using the same accelerator) jointly announced that all doubt had disappeared. Thirty-eight photographs of top quark collisions had been taken. The top quark was finally snared.

Generations of Quarks

To understand the importance of finding the top quark, we need to know that quarks come in several pairs, or "generations." The lowest pair are called the "up" and "down" quarks. When three of these light quarks are combined together, we find the familiar protons and neutrons that make up the visible universe, including the atoms and molecules in our own bodies. (Three quarks make up the protons and neutrons. The proton, for example, consists of two up quarks and one down quark, while the neutron consists of two down quarks and one up quark.) Each up and down quark, in turn, comes in three different "colors," making up a total of six quarks for the first generation. (This "color" has nothing to do with the familiar concept of color.)

The next heavier pair of quarks are called the "strange" and "charmed" quarks. When they are joined together, they form many of the heavy fragments that are found among the debris created by smashing atoms apart. These quarks, not surprisingly, also come in three colors.

One of the deepest mysteries of matter (which even today is not understood) is why the first and second generation of quarks seem to be almost carbon copies of each other. Except for the fact that the second pair is heavier than the first pair, they have almost identical properties. It seems strange that nature, at a fundamental level, prefers to have a high degree of useless redundancy in the construction of the universe.

The discovery of the bottom quark in 1977 meant there had to be a third redundant generation of quarks and also a missing top quark to fill up the third pair. The foundation of the Standard Model was therefore based on three generations of quarks, each generation being identical to the previous generation, except for its mass.

Today, physicists say that quarks come in six "flavors" (up, down, strange, charm, bottom, and top) as well as three colors. This makes eighteen quarks. But each quark also has an antimatter twin. When we add in the antiquarks, the total number of quarks comes to thirty-six. (This number is much larger than the total number of subatomic particles found in the 1930s, when many physicists thought that the electron, proton, and neutron were enough to describe all the matter in the universe.)

The Standard Model

At present, there has been no experimental deviation from the Standard Model. Thus, it is perhaps the most successful theory ever proposed in the history of science. However, most physicists find the Standard Model unappealing because it is exceptionally ugly and asymmetrical. (For a more detailed discussion of symmetry in physics, see chapter 7.) Because experimentally it is very successful, most physicists believe that the Standard Model is just an intermediate step toward a true theory of everything. The reason why the Standard Model is so ugly is that it is obtained by gluing, by brute force, the current theories of the electromagnetic force, the weak force, and the strong force into one theory. (Think of trying to jam three jigsaw puzzle pieces together that obviously do not fit. Tape can be used to connect the three pieces forcibly. That is the Standard Model.)

To understand how ugly the theory is, let us summarize how the various pieces fit together.

First, the strong interactions are described by these thirty-six quarks, coming in six flavors, three colors, and matter/antimatter pairs. The "glue" that holds them together to form the proton and neutron are the gluons (which are described by the Yang-Mills field). Altogether, there are eight gluon fields. Collectively, this theory is called quantum chromodynamics, or the theory of the "color" interactions.

The weak interaction also has the same generation problem. The first generation has the electron and the neutrino. The second generation has the muon and its neutrino. The third generation has the tau particle and its neutrino. Collectively, these particles are called "leptons," and they are the counterparts of the quarks that are found in the strong interactions. These leptons, in turn, interact by exchanging the W and Z particles (which are massive Yang-Mills fields). Altogether, there are four such particles.

Then there are the electromagnetic interactions, which are mediated by the Maxwell field.

And last, there is something called the "Higgs particle" (a particle that allows us to break the symmetry of the Yang-Mills field). All these particles except for the Higgs particle have been discovered in atom smashers.

At present, physicists have probed the interactions of subatomic particles past a trillion electron volts and have found no experimental deviation from the Standard Model. However, although the theory is undeniably successful, it is unappealing; we know it cannot be the final theory because:

1. It has such a bizarre collection of quarks, leptons, gluons, and W, Z bosons.

2. It has exactly three generations, in both the quark and lepton sector, which are indistinguishable (except for their masses).

3. It has nineteen arbitrary parameters, including the mass of the leptons, the mass of the W and Z particles, the relative strength of the strong and weak interactions, and so on. (The Standard Model does not determine the value of these nineteen numbers. They are inserted ad hoc in the model, without justification, and are fixed by carefully measuring the properties of these particles.)

As a guiding principle, Einstein would always ask himself the question: If you were God, how would you construct the universe? Certainly not with nineteen adjustable parameters and a horde of redundant particles. Ideally, you would only want one (or no) adjustable parameters, and just one object out of which to construct all the particles found in nature, and maybe even space and time.

By analogy, we can look at the Mendeleev chart, with its collection of over one hundred elements, which were the ''elementary particles'' of the last century. No one could deny that the Mendeleev chart was spectacularly successful in describing the building blocks of matter. But the fact that it was so arbitrary, with hundreds of arbitrary constants, was unappealing. Today, we know that this entire chart can be explained by just three particles, the neutron, the proton, and the electron. Similarly, physicists believe that the Standard Model, with its odd-looking and redundant quarks and leptons, should be constructed from even simpler structures.

One intermediate step would be to combine the various Yang-Mills fields into one unified set of fields.

GUTs and Renormalization

The simplest theory that would reshuffle these particles into one another was called SU(5), written down by Howard Georgi and Sheldon Glashow of Harvard in 1974. In this ''grand unified theory'' (GUT), the symmetry SU(5) linked the electron, the neutrino, and the quarks. Correspondingly, the photon, the W-particle of weak interactions, and the gluons of the strong interactions were now being pieced together to form another family of forces.

Although the GUT theory is difficult to test, because the energy at which strong interactions are unified with the electro-weak force is beyond the range of our present particle accelerators, the GUT theory does make a remarkable prediction that can be tested with today's technology.

This theory predicts that the quark can turn into an electron by emitting another particle. This means, of course, that the proton (made up of three quarks) will eventually decay into electrons and that the proton has a finite lifetime. This remarkable prediction of the GUT theory—that the proton eventually should decay into electrons—has prompted a new generation of experimental physicists around the world to work at testing it. (Although several groups of experimental physicists are looking for evidence of proton decay with detectors buried deep in the earth, no one yet has conclusively identified proton decay.)

In retrospect, although the GUT theory represented a remarkable advance in unifying the electro-weak force with the strong force, it had grave experimental problems. For example, outside of proton decay experiments, it is very difficult if not impossible to test the GUT predictions directly.

More important, the GUT theory is also incomplete theoretically. It does not explain, for example, why there are three carbon-copy families of particles (the electron, muon, and tau families). Furthermore, scores of arbitrary constants (such as the masses of quarks, the masses of leptons, and the number of Higgs particles) are littered throughout the theory. After a while, with so many undetermined parameters, the GUT theory seems to resemble a Rube Goldberg device. To a physicist, it is hard to believe that a theory can be fundamental with so many outstanding parameters.[3]

Despite the problems of the GUT theory, however, physicists were still hopeful that lightning might strike five times. Would a simple gauge theory (such as Yang-Mills) yield a theory of gravity?

The answer was a resounding no. Gauge theory, for all its successes, hit a brick wall when dealing with gravity. The Yang-Mills formalism was still too primitive to account for gravity. This is perhaps the most fundamental objection to the GUT theory; despite its successes, the theory is incapable of including gravitational interactions.

No progress would be made in this area until the birth of a new idea, based on symmetries even larger than the Yang-Mills theory. That theory would be the superstring theory.

II

Supersymmetry and Superstrings

6

The Birth of the Superstring Theory

THE SUPERSTRING THEORY has perhaps the weirdest history in the annals of science. Nowhere else do we find a theory that was proposed as the solution to the wrong problem, abandoned for over a decade, and then resurrected as a theory of the universe.

The superstring theory began in the 1960s, before the flourishing of the Yang-Mills theory and gauge symmetries, when the renormalization theory was still floundering as a theory bedeviled by infinities.

A backlash had developed against the renormalization theory, which seemed contrived and artificial. The opposing school of thought was led by Geoffrey Chew of the University of California at Berkeley, who proposed a new theory that was independent of elementary particles, Feynman diagrams, and the renormalization theory.

Instead of postulating a series of intricate rules detailing how certain elementary particles interact with other particles through Feynman diagrams, Chew's theory required only that the S-matrix (which mathematically describes the collisions of particles) be self-consistent. Chew's theory postulated that the S-matrix obeys a rigorous set of mathematical properties, and then assumed that these properties are so restrictive that only one solution was possible. This approach is often called the "bootstrap" approach, because one is literally picking oneself up by one's bootstraps (one begins with only a set of postulates, then theoretically derives the answer using only self-consistency).

Because Chew's approach was based entirely on the S-matrix, rather than on elementary particles or Feynman diagrams, the theory was called the "S-matrix theory" (not to be confused with the S-matrix itself, which all physicists use).

These two theories, quantum field theory and S-matrix theory, are based on different assumptions about the meaning of an "elementary particle." The quantum field theory is based on the assumption that all matter can be built from a small set of elementary particles, whereas the S-matrix theory is based on an infinite number of particles, with none of them elementary.

In retrospect, we see that the superstring theory combines the best features of the S-matrix theory and the quantum field theory, which in many ways are opposites.

The superstring theory resembles the quantum field theory because it is based on elementary units of matter. Instead of point particles, however, the superstring theory is based on strings that interact by breaking and reforming via Feynman-like diagrams. But the significant advantage that superstrings have over the quantum field theory is that renormalization is not required. All the loop diagrams at each level are probably finite by themselves, requiring no artificial sleights of hand to remove the infinities.

Similarly, the superstring theory resembles the S-matrix theory in that it can accommodate an infinite number of "elementary particles." According to this theory, the infinite variety of particles found in nature are simply different resonances of the same string, with no particle any more fundamental than any other. The great advantage, however, that the superstring theory has over the S-matrix theory is that it is possible to calculate with the superstring theory and eventually get numbers for the S-matrix. (By contrast, the S-matrix theory is exceedingly difficult to calculate with and extract usable numbers.)

The superstring theory, then, incorporates the best features of both the S-matrix theory and the quantum field theory because it is based on a startlingly different physical picture.

The superstring theory, unlike the S-matrix theory or the quantum field theory, which were based on years of patient development, burst forth unexpectedly on the physics community in 1968. In fact, it was by sheer accident, and not a logical sequence of ideas, that the superstring idea was discovered.

Guessing the Answer

In 1968, when the S-matrix theory was still very much in vogue, two young physicists, Gabriele Veneziano and Mahiko Suzuki, each working independently at CERN, the nuclear research center outside Geneva, asked themselves a simple question: If the S-matrix is supposed to obey so many restrictive properties, then why not just try to guess the answer? They thumbed through voluminous tables of mathematical functions cataloged since the eighteenth century by mathematicians and stumbled upon the Beta function, a beautiful mathematical formula first written down by the Swiss mathematician Leonhard Euler in the 1800s. Much to their astonishment, upon examining the properties of the Beta function, they found that it automatically satisfied almost all of Chew's S-matrix postulates.[1]

This was crazy. Was the solution to strong interaction physics simply a formula written down more than a hundred years earlier by a mathematician? Was it all so simple?

Making a major scientific discovery by randomly flipping through a math book had never happened before in the history of science. (Perhaps the fact that Veneziano and Suzuki were both too young to appreciate the odds against their random discovery helped them to find the Beta function. An older, more prejudiced physicist might have dismissed from the start the idea of finding the answer in an old math book.)

Euler's formula became an overnight sensation in the world of physics—the apparent victory of the S-matrix theory over the quantum field theory. Hundreds of papers were written trying to use the Beta function to fit the data pouring out of atom smashers. Many papers, in particular, were written to solve the last remaining postulate of Chew's that the Beta function did not obey: unitarity, or the conservation of probability.

Very rapidly, attempts were made to propose even more complex theories that would fit the data even better. Soon John Schwarz and French physicist Andre Neveu, both working at Princeton University at the time, and Pierre Ramond, then at the National Accelerator Laboratory near Chicago, proposed a theory that would include particles with "spin" (which eventually became the superstring theory).

As remarkable as the Beta function was, the nagging question remained: Were the marvelous properties of this formula strictly an accident, or did they arise from a deeper, more physical underlying structure? The answer was finally established in 1970, when Yoichiro Nambu of the University of Chicago showed that this marvelous Beta function was due to the properties of interacting strings. When this new approach was applied to the Neveu-Schwarz-Ramond theory, it became the present theory of the superstring.

The Nambu Mode

Unlike Einstein, who delighted in thumbing his nose at pompous social formalities, or Feynman, who loved pranks, or Gell-Mann, the *enfant terrible* of physics, Nambu is renowned for his quiet, well-mannered, but always penetrating style. He has much of the character of the traditional Japanese, who are more reserved, some would say more thoughtful, in their style than their often brash Western colleagues. In the rough-and-tumble marketplace of ideas, where credit for originating certain physical ideas is jealously guarded, Nambu has a refreshingly different style, preferring to let the merit of his work speak for itself.

This means, however, that although he has participated in some of the most basic discoveries in physics, he has not insisted on claiming priority. In physics, names are often attached to discoveries by general consensus, even though that may not be precisely correct historically. For example, the well-known Bethe-Saltpeter equation, which describes the behavior of two electron systems, was first published by Nambu. Similarly, Nambu first published many of the early ideas of "spontaneous symmetry breakdown," although for years it was known as the "Goldstone" theorem. Only recently has it been properly called the Nambu-Goldstone theorem. However, with superstrings, it was clearly Nambu who wrote down the basic equations of the string theory.[2]

One reason why some of his remarkable achievements have not won instant recognition is that he is usually far ahead of his time. As his colleague Dr. Laurie Brown of Northwestern University has noted, Nambu is a "trailblazer whose innovations set the stage for breakthroughs, typically years or even decades before their realization by others."[3] It's a saying among physicists that if you want to know what physics will be like in the next decade, you read the works of Nambu.

At a talk in 1985, Nambu tried to sum up the modes of thinking used by the great physicists of the past that led to pathbreaking discoveries. Nambu called them the "Yukawa mode" and the "Dirac mode" of thinking. The Yukawa mode is deeply rooted in experimental data. Yukawa was led to his seminal idea of the meson as the carrier of the nuclear force by closely analyzing the data available to him. The Dirac mode, however, is the wild, speculative leap in mathematical logic that led to astonishing discoveries, such as Dirac's theory of antimatter or his theory of the monopole (a particle that represents a single pole of magnetism). Einstein's theory of general relativity would fit into the Dirac mode.

However, at a celebration of Nambu's sixty-fifth birthday in 1985, in which his vast scientific achievements were summarized, his colleagues coined in his honor yet another way of thinking: the "Nambu mode." This mode combines the best features of both modes of thinking and tries carefully to interpret the experimental data by proposing imaginative, brilliant, and even wild mathematics. The superstring theory owes much of its origin to the Nambu mode of thinking.

Perhaps some of Nambu's style can be traced to the clash of Eastern and Western influences represented by his grandfather and father. After the disastrous 1923 earthquake that leveled Tokyo, Nambu's family settled in the small town of Fukui, well known as the seat of the Shin-shu sect of Buddhism. Nambu's grandfather supported the family by selling religious objects, such as household shrines for honoring one's ancestors. Nambu's father, instead of dutifully following in his father's ways, rebelled and ran away from home several times. An intellectual, Nambu's father was fascinated by Western culture and eventually was graduated with a major in English literature, writing his thesis on William Blake.

Nambu grew up in this household, dominated by a traditionalist grandfather but tempered by the strange intellectual winds blowing from the West. However, the entire family suffered when militarism rose in Japan in the 1930s. As Dr. Brown noted, Nambu's father

> had liberal and internationalist views, which it was politically prudent in those days to keep to oneself. He subscribed to several series of inexpensive books (so-called yen books), which Yoichiro read. These included foreign novels, modern Japanese literature, and the Marxist classics. The latter continued to arrive even during the 1930s, but they began to be heavily censored. Finally, it became dangerous to own such books, but Nambu's father kept some of them.[4]

As a child Nambu showed an interest in science and, like Feynman and many others, tinkered with small radio receivers. While a student at the University of Tokyo he was fascinated by stories of the new quantum mechanics being developed in the West by Heisenberg and others. However, Nambu detested the militaristic atmosphere that was gripping the country.

After the disastrous defeat in 1945, the Japanese people began the painful process of rebuilding their nation. Nambu had an appointment at the University of Tokyo, where Japanese physicists, such as Shinichiro Tomonaga, cut off by the war from the work of their counterparts in the West, slowly began to restore their international contacts.

Princeton physicist Freeman Dyson captured the pleasant surprise that physicists in the West felt upon receiving news of progress in Japan when he wrote that Tomonaga

> set out simply and lucidly without any mathematical elaboration, the central idea of Julian Schwinger's theory. The implications of this were astonishing. Somehow or other, amid the ruin and turmoil of the war, totally isolated from the rest of the world, Tomonaga had maintained in Japan a school of research in theoretical physics that was in some respects ahead of anything existing anywhere else at that time. He had pushed on alone and

laid the foundations of the new quantum electrodynamics, five years before Schwinger. . . .[5]

Nambu's work eventually caught the attention of J. Robert Oppenheimer, director of the Institute for Advanced Studies at Princeton, who invited him to stay at the institute for two years. Nambu left Japan in 1952, and was shocked upon encountering a "normal" society. (Tokyo, because of the massive firebombing raids, received more damage than Hiroshima.) In 1954, he visited the University of Chicago, where he has been a professor since 1958.

The sharp contrast between Nambu's soft, reserved style and Feynman's outspoken manner was graphically displayed in 1957 at the Rochester Conference in Rochester, New York, when Nambu presented a paper postulating the existence of a new particle or resonance (the isoscalar meson). When Nambu gave his talk, Feynman responded by shouting "In a pig's eye!" (A few years later, however, the question was settled when this particle was discovered in atom smashers and christened the "omega meson.")

Nambu's String

Nambu originally proposed the idea of the string to make some sense out of the chaos of the hundreds of hadrons being discovered in the nation's laboratories. Clearly, these hadrons could not be viewed as "fundamental" in any sense of the word. The disarray of strong interaction physics, Nambu thought, must be a reflection of some underlying structure.

One proposal, made years earlier by his colleague Yukawa, and others such as Heisenberg, assumed that elementary particles were not points at all but "blobs" that could pulsate and vibrate. Over the years, all efforts to build a quantum field theory based on blobs, membranes, and other geometric objects have failed. These theories eventually violated some physical principle, such as relativity (because if the blob was shaken at one point, the vibration could travel through the blob faster than the speed of light). They were only vaguely defined and extraordinarily difficult to use in any calculation.

Nambu's seminal idea was to assume that the hadron consisted of a vibrating string, with each mode of vibration corresponding to a separate particle. (The superstring theory would not violate relativity because vibrations along the string could travel only less than or at the speed of light.)

Think of our previous analogy of the violin string. Let's say that we are given a mysterious box that creates musical tones. If we knew nothing about music, we would first attempt to catalog the musical tones, giving them names, such as C, F, G, and so on. Our second strategy would be to discover relations among the notes, such as observing that they occur in groups of eight (octaves). From this we would be able to discover the laws of harmony. Last, we would try to postulate a "model" that would explain the harmonies and musical scale from a single principle, such as a vibrating violin string. Similarly, Nambu believed that the Beta function found by Veneziano and Suzuki could be explained by vibrating strings.

One problem remaining was to explain what happened when strings collided. Because each mode of the string represents a particle, understanding how strings collide allows us to calculate the S-matrix of ordinary particle interactions. Three physicists then working at the University of Wisconsin, Bunji Sakita, Keiji Kikkawa, and Miguel Virasoro, conjectured that the last remaining postulate (unitarity) of Chew's S-matrix could be satisfied in the same way as the renormalization theory solves this postulate: by adding loops. In other words, these physicists proposed to reintroduce Feynman diagrams for these strings. (At this point, many of the S-matrix theorists were dismayed. This heretical idea meant reintroducing loops and the renormalization theory, which they had banned from the S-matrix theory. This was too much for the purists in the S-matrix camp.)

Their proposal was finally completed by one of us (Michio) and a collaborator, Loh-Ping Yu, when they were graduate students at the University of California at Berkeley, and also by Claude Lovelace, then at CERN, and V. Alessandrini, a physicist from Argentina.

Fun with Origami

Strings come in two types: open strings (which have ends) and closed strings (which are circular). To understand how strings interact, think of the Tinker Toys that represent Feynman diagrams for point particles. When a particle moves, it creates a line, which is represented by a Tinker Toy stick. When particles collide, they form Y-shaped lines, with the collision represented by a Tinker Toy joint.

Similarly, when open strings move, their paths can be visualized as long paper strips. When closed (circular) strings move, their paths can be visualized as paper tubes, not lines. Therefore, we need to replace Tinker Toys with origami.

A

B

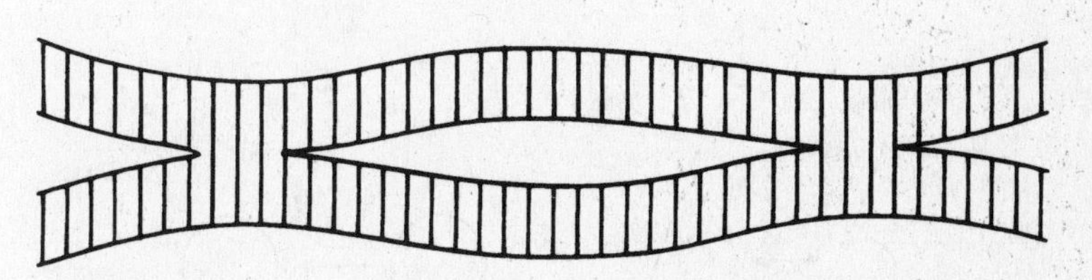

In diagram A, two closed strings enter from the left, collide in the middle, and form a single string, which then breaks in half and forms two strings. In diagram B, two open strings enter from the left, merge, break, merge, and break again into two strings that go off to the right.

When these paper strips collide, they merge smoothly to form another paper strip. Again, we have a Y-shaped joint, but the lines forming the Y are strips, not sticks.

This means that physicists, instead of doodling with lines on a blackboard, must visualize colliding paper strips and tubes. (Michio can remember one conversation with his adviser, Stanley Mandelstam of Berkeley, who used scissors, tape, and paper to explain how two strings can collide, re-form, and create new strings. This paper construction eventually evolved into an important Feynman diagram for superstrings.)

When two strings collide and produce an S-matrix, we use Feynman diagrams shown on the opposite page.

The field theory of these interactions was completed by Michio and Keiji Kikkawa in 1974. They showed that the entire superstring theory could be summarized as a quantum field theory based on strings, not point particles. Only five types of interactions (or joints) were needed to describe string theory:

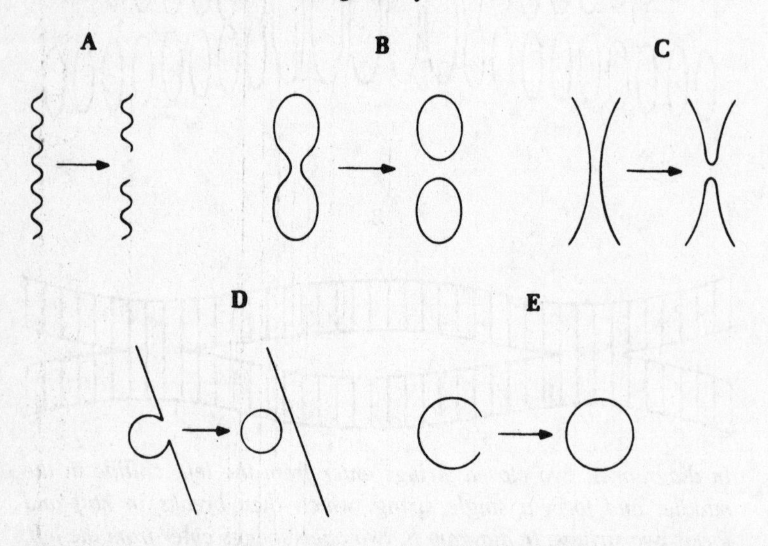

The five types of string interactions are represented here. In diagram A, a string splits and creates two smaller strings; in diagram B, a closed string pinches and creates two smaller strings; in diagram C, two strings collide and re-form into two new strings; in diagram D, a single open string re-forms and creates an open and a closed string; in diagram E, the ends of an open string touch and create a closed string.

The test of the theory is when we generalize these Feynman diagrams to "loops." As before, all the divergences (if there are any) in the Feynman diagrams occur when the string forms loops. In ordinary renormalization theory, we are allowed to reshuffle these divergences and use other gimmicks in order to eliminate them. In any gravity theory, however, this reshuffling is impossible, and each term in the series must be finite. This places tremendous restrictions on the theory. A single infinite Feynman diagram can spoil the entire program. As a result, for decades physicists despaired of ever being able to eliminate these infinities.

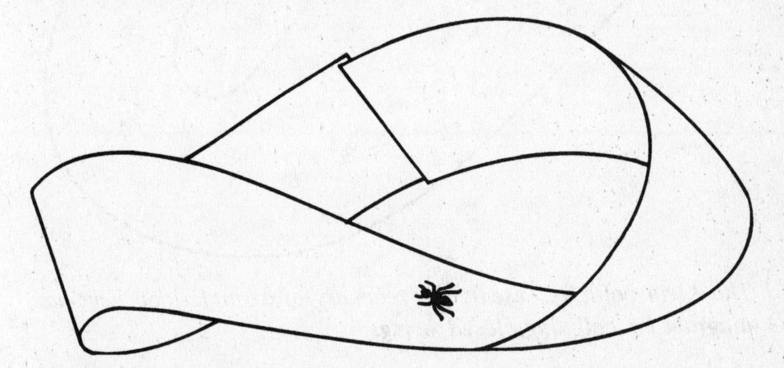

The Möbius strip represents the geometry of a single-loop Feynman diagram for colliding open strings.

Astonishingly, the Feynman diagrams for interacting string are known to be finite. A marvelous sequence of cancellations occurs that seems to eliminate all potentially infinite terms, yielding a finite answer.

Proving the lack of divergence of the superstring theory requires some of the most bizarre constructions of geometry. For example, in a simple one-loop diagram, the interior of the Feynman diagram is represented by a circular strip or tube.

The complete theory, however, demands that the paper strip or tube be twisted. If we twist a circular strip, we arrive at a geometric object called the Möbius strip (which is a strip with only one side). Everyone knows that a strip of paper has two sides to it. However, if we twist one side and then glue the two sides together, we have a

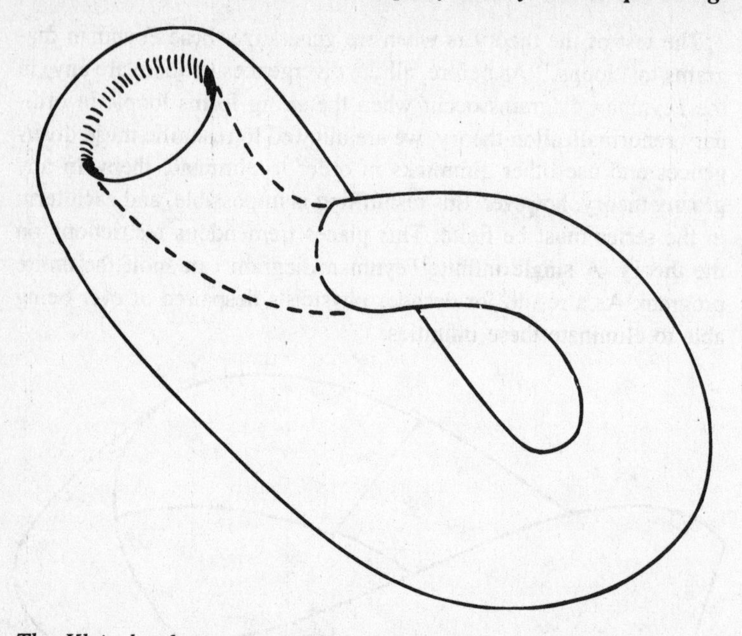

The Klein bottle represents the geometry of a single-loop Feynman diagram for colliding closed strings.

one-sided strip. An ant walking along the inside of this strip soon finds itself walking along the outside. Similarly, when twisting a circular tube, we arrive at an even more bizarre object called the Klein bottle, a two-dimensional surface with only one side. Everyone knows that a hollow tube has two sides—an inside and an outside. However, if we twist one end of the tube 180 degrees, and then distort the tube by joining these two ends, we arrive at a Klein bottle.

Historically, the Möbius strip and the Klein bottle were little more than geometric curiosities, with no practical applications. To a string physicist, however, both appear as part of the Feynman diagram containing loops and are essential for the cancellation of divergences.

The Death of the Superstring Theory

Although the superstring theory was a beautiful mathematical abstraction that seemed to fit some of the strong interaction data, there were frustrating difficulties with the model.

First, the theory predicted too many particles. The theory had particles that acted like "gravitons" (quantum packets of the gravitational force) and photons (packets of light). In fact, the lowest vibration of the closed string corresponded to the graviton and the lowest vibration of the open string corresponded to the photon.

This was disastrous for a theory that was supposed to describe the strong interactions, not gravity or electromagnetism. What were the graviton and the photon doing in a theory of strong interactions? (The fact that this was a blessing in disguise was not appreciated at the time. The gravitation and light interactions occurring in the string model are precisely what is necessary to make a unified field theory.)

Second, the theory seemed to predict the existence of "tachyons," which are particles that go faster than the speed of light. These particles were undesirable, because they implied that you could violate causality—that is, go back in time and meet your mother before you were born.

Third, and most devastating, physicists soon discovered that the original Nambu theory was self-consistent only in twenty-six dimensions. (For any theory, being inconsistent is the kiss of death. For example, if a theory is inconsistent, it will eventually make ridiculous predictions, such as $1 + 1 = 3$.)

Claude Lovelace at CERN first discovered that the string model seemed to have a better mathematical structure in twenty-six dimensions. Then Richard Brower and Charles Thorn at MIT and others showed that the model collapsed unless the theory was defined in twenty-six dimensions. Soon physicists discovered that the superstring theory (the Neveu-Schwarz-Ramond model) was self-consistent only in ten dimensions.

This was simply too much for most physicists. For scientists used to thinking in four dimensions, this theory sounded more like science fiction than true science. As a result, the superstring theory lost favor around 1974. Many physicists, Michio included, reluctantly deserted the model.

Michio can still remember the shock and dismay that many physicists felt knowing that the model was only consistent in twenty-six and ten dimensions. We all remembered Niels Bohr's quote, that any great theory should be "crazy enough," but it stretched the limits of our scientific imaginations to believe that the universe could be in twenty-six or even ten dimensions.

Space, as everyone knows, has three dimensions: length, depth, and breadth. The size of any object in our universe—anything from an ant to the sun—can be described in terms of these three quantities.

If we want to describe, say, the age of the sun, we need one more quantity: time. With these four quantities (length, depth, breadth, and time), we can describe the physical state of any object in the universe. Consequently, physicists say that we live in a four-dimensional universe.

A favorite device of science fiction writers has been to invent more than four dimensions, to assume that "parallel universes" exist, similar to our own but in different dimensions. But it was merely a writer's device; physicists never took the idea of parallel universes seriously. So when the string model predicted a higher-dimensional universe, it was dismissed by most physicists.

The decade of 1974 to 1984 was lean for the string model, with most physicists working on the fast-paced developments in electroweak and GUT-type theories. Only the most dedicated workers, such as Michael Green of Queen Mary College in London and John Schwarz of the California Institute of Technology, puttered with the theory.

In 1976, several physicists tried to revive the theory by making an outlandish suggestion. Joel Scherk in Paris and John Schwarz suggested that the string model be reinterpreted. They decided to make a vice into a virtue. Perhaps the theory's unwanted "graviton" and "photon" were the real graviton and photon after all. In their approach, the superstring theory was the correct theory being used for the wrong problem. Instead of a strong interaction theory, it was actually a theory of the universe!

This reinterpretation of the string model was met with extreme skepticism. After all, the theory was only moderately successful at predicting strong interactions, and now Scherk and Schwarz were making it into a theory to explain the universe. This idea, although clever, was not taken seriously. After all, the theory was still in ten dimensions. Schwarz summed up the situation when he said, "No one accused us of being crackpots, but our work was ignored."[6]

Children of the String

Ironically, although the superstring theory died in the 1970s as a model of strong interactions, the next decade witnessed the flowering of what we might call the "children of the string." The superstring theory was considered too symmetrical to be realistic, so other theories that corporated certain features of the model became fashionable. Although the string itself fell into disfavor, many of its byproducts dominated and cross-pollinated theoretical physics during that decade from 1974 to 1984. The string had such a rich theoretical structure that its spinoffs circulated within the physics community.

For example, Ken Wilson of Cornell University used the novel concept of a string to propose that quarks were permanently bound together by a stringlike sticky substance. He proposed this theory to answer a puzzling question: Where were the quarks? Although quarks had been universally accepted by the physics community for the past twenty years, no one had seen one in the laboratory. Gell-Mann and others proposed that somehow these quarks could be "confined" by a mysterious force.

Wilson's theory proposed that the Yang-Mills gluons found in the quark theory, which normally occur as particles, could, under some conditions, "condense" into a sticky kind of taffy that would confine the quarks. Wilson used computers to show that these gluon particles could condense into taffylike strings with quarks at either ends, much the same way as steam can condense into water droplets. Quarks were never seen, according to this logic, because they are permanently confined by strings.

Today, the National Science Foundation is allocating millions of dollars to build some of the largest computers in the world (called fifth-generation computers) to answer questions like the one posed by Wilson. Wilson's string theory, in principle, is powerful enough to calculate virtually all the properties of strong interactions. For his pioneering work in this field, called ''phase transitions,'' which has an immediate impact on solid state physics and the quark model, Wilson was awarded the Nobel Prize in 1983.

Another spinoff of the string was ''supersymmetry'' (to be discussed in a later chapter). Although supersymmetry was first found in a ten-dimensional theory, it also could be applied to four-dimensional theories, and it became quite fashionable by the late 1970s. GUTs, it turned out, suffered from certain diseases that supersymmetry could cure.

Later, a more sophisticated version of supersymmetry, one that included gravity, was proposed, called ''supergravity.'' This theory, first formulated by Peter van Nieuwenhuizen, Dan Freedman, and Sergio Ferrara, then at the State University of New York at Stony Brook, became the first nontrivial extension of Einstein's equations in sixty years. (The supergravity theory, because it is based on supersymmetry, actually is contained within the superstring theory.)

Finally, even the prejudice of physicists against higher dimensions of space-time began to break down in the early 1980s, when Kaluza-Klein models became fashionable. Certain quantum effects could make even higher-dimensional theories physically acceptable. (This will be explained in greater detail later.)

Although the children of the string dominated the direction of theoretical physics in the late 1970s and early 1980s, the parent was largely shunned. Here was a theory that had the largest set of symmetries known to science, yet it was considered totally useless. This began to change, however, in 1984, when physicists reexamined something called ''anomalies.''

Triumph of Accident and Shrewd Observation

Anomalies are another by-product of marrying quantum mechanics with relativity. Anomalies are tiny but potentially deadly mathematical defects in a quantum field theory that must be canceled or eliminated. A theory just doesn't make sense in the presence of these anomalies.

Anomalies are like the small defects that occur when blending the finest clay, sand, and minerals in order to make glazed pottery or ceramics. If even a slight mistake is made in mixing the right proportion of ingredients, this small blemish can ruin the finished product, causing it eventually to crack.

Anomalies tell us that the theory, no matter how elegant, is ultimately inconsistent and will make ridiculous predictions. Anomalies also tell us that nature requires yet another constraint in building a quantum field theory of gravity. In fact, there seem to be so many constraints on the quantum theory that, as with the S-matrix theory, the final answer may be unique.

Anomalies exist in most theories with symmetry. For example, the superstring model is in ten dimensions (as the Russian physicist A. M. Polyakov showed) because a higher dimension is required to eliminate an anomaly.

Edward Witten and Luis Alvarez-Gaume of Princeton University found that when quantum field theory is used to describe gravity interacting with other particles, the theory is riddled with fatal anomalies. Then in 1984 Green and Schwarz observed that the superstring model possesses enough symmetry to ban anomalies once and for all. The symmetry of the superstring, once considered too beautiful to have any practical application, now became the key to eliminating all infinities and anomalies.

This realization touched off an explosion of interest in the superstring theory. Nobel laureate Steven Weinberg, hearing of the excitement over superstrings, immediately switched to working on the superstring theory. "I dropped everything I was doing," he recalls, "including several books I was working on, and started learning everything I could about string theory." It wasn't easy, however, to learn an entirely new mathematics. "The mathematics is very difficult,"[7] he concedes.

The transformation was startling. Within a few months the superstring theory went from little more than a beautiful but useless curiosity to perhaps the only hope for a unified field theory. Anomalies, instead of destroying any hope of building a quantum theory of gravity, resurrected the superstring. The number of superstring papers being published—merely a trickle in the early 1980s—grew to over a thousand by 1995, making the theory a dominant force in theoretical physics.

There are a few other rare occasions in the history of science when an apparent defect was found to be a tremendous asset. For example, in 1928, Alexander Fleming found that his cultured dishes of staphylococcus bacteria colonies could be destroyed if they were accidentally contaminated by certain bread molds. At first he found it a nuisance to take protective measures to prevent the bacteria cultures from being damaged by these molds. But then it dawned on Fleming that perhaps the bacteria-killing molds were more important than the bacteria cultures themselves. This observation led to the discovery of penicillin, which led to Fleming's receiving the Nobel Prize in medicine in 1945 for what he called the "triumph of accident and shrewd observation."

Like a phoenix rising from its own ashes, the superstring theory returned, this time with a vengeance, thanks largely to Schwarz and Green's triumph of accident and shrewd observation.

7

Symmetry: The Missing Link

WHAT IS BEAUTY?

To a musician, beauty might be a harmonious, symphonic piece that can stir up great passion. To an artist, beauty might be a painting that captures the essence of a scene from nature or symbolizes a romantic concept. To a physicist, however, beauty means symmetry.

In physics, the most obvious example of symmetry is a crystal or gem. Crystals and gems are beautiful because they have symmetry—they retain the same shape if we rotate them at certain angles.

We say that the crystal is *invariant* under a rotation at a certain angle, because the crystal rotates back into itself. A cube, for example, retains its original orientation if we rotate it by ninety degrees around any of its axes. A sphere has even more symmetry because it is invariant under all possible rotations because it rotates into itself.

In much the same way, when we apply symmetry to physics, we demand that the equations remain the same when we make certain "rotations." In this case, rotations (actually, shufflings) occur when we change space into time, or electrons into quarks. We say our equations retain a beautiful symmetry if, after making these rotations, the equations remain the same.

Physicists have often debated the question: Is symmetry simply a matter of aesthetics, peculiar only to humans, or does nature prefer symmetry in the universe?

The universe certainly is not created symmetrically. The universe does not consist entirely of beautiful ice crystals and gems, but instead appears horribly broken. There are not many symmetries left in jagged rocks, meandering rivers, formless clouds, irregular mountain ranges, random chemical molecules, or the blizzard of known subatomic particles.

With the discoveries made with the Yang-Mills and gauge theories, however, we are beginning to realize that nature, at the fundamental level, does not just prefer symmetry in a physical theory, nature demands it. Physicists now realize that symmetry is the key to constructing physical laws without disastrous anomalies and divergences.

Symmetry explains why all the potentially harmful divergences and anomalies, sufficient to kill other theories, cancel each other perfectly in the superstring theory. The superstring model, in fact, has such a huge set of symmetries that the theory can include all the symmetries of the electro-weak and the GUT-type theories, as well as Einstein's theory of general relativity. All the known symmetries of the universe, and many that have not yet been discovered, are found within the superstring theory. In retrospect, it is clear that symmetries are the reason why the superstring theory works so well.

Physicists now appreciate the fact that symmetries are essential to eliminate potentially fatal problems that face any relativistic quantum theory. Although scientists prefer that a theory possess symmetry for purely aesthetic reasons, they are learning that nature actually demands symmetry from the start as an ironclad criterion for an acceptable merger of relativity and quantum mechanics.

This is not obvious from the start. Previously, physicists believed that they could write down many possible self-consistent theories of the universe that were relativistic and obeyed quantum mechanics. Now, to our surprise, we are finding that perhaps the conditions for the elimination of divergences and anomalies are so stringent that only one theory is allowed.

Symmetry and Group Theory

The mathematical study of symmetry is called "group theory" (where a group is simply a set of mathematical objects linked by precise mathematical rules), which owes its origin to the work of the great French mathematician Evariste Galois, who was born in 1811. Using the power of symmetry alone, Galois as a teenager solved a problem that had stumped the world's greatest mathematicians for five hundred years. For example, if we have the equation $x^2 + bx + c = 0$, we are taught in high school algebra that we can find a solution for x using only square roots. The question was: Can the quintic (fifth power) equation, $ax^5 + bx^4 + cx^3 + dx^2 + ex + f = 0$, also be solved in such a manner?

Amazingly, this teenager created a new theory that was so powerful that it could answer this problem that had eluded the best minds in the world of mathematics for centuries. The answer was no. His solution demonstrated the enormous power of group theory.

Unfortunately, Galois was so far ahead of his time that other mathematicians did not appreciate his path-breaking research. When he applied for entrance into the prestigious École Polytechnique, for example, he gave a lecture on mathematics that soared above the heads of the examination committee. As a consequence, he was rejected.

Galois then summarized his key discoveries and sent the paper to the mathematician Augustin-Louis Cauchy for presentation before the French Académie. Cauchy, who failed to realize the importance of this work, subsequently lost Galois's paper. In 1830, Galois submitted another paper to the Académie in competition for its prize, but this time the referee, Joseph Fourier, died shortly before the competition and the paper was lost. Frustrated, Galois submitted his paper to the Académie one last time, but this time the mathematician Simeon-Denis Poisson dismissed it as "incomprehensible."

Galois was born into a world where revolution was sweeping the land, and he embraced the causes of the Revolution of 1830. He was finally accepted for admission to the École Normale in Paris, but he was soon expelled because he was a radical. He was arrested in 1831 for agitation at a rally against King Louis Philippe. History records that a year later a police agent, an agent provocateur, set up a duel with him. (Galois apparently was involved with a woman and was bound by a code of honor to duel with pistols.) Galois was killed when he was barely twenty years old.

Fortunately, the evening before the duel, Galois had a premonition of his death. He wrote down his key results in a letter to his friend Auguste Chevalier and asked that it be published in the *Revue Encyclopedique.* This letter, which contains his key ideas on group theory, was not published for fourteen years. (A century later, mathematicians still puzzle over his notes, because he made references to mathematical equations that were not discovered until twenty-five years after his death.)

Although group theory suffered an indisputable loss with the death of its founder, Galois, the point here is to show the enormous power contained within the theory. Not only was it mathematically elegant in its own right, it possessed tremendous firepower when applied to other mathematical problems. There is something strange and wondrous about a symmetry that enables us to solve problems that cannot be solved by any other means. (Group theory is now so much a part of mathematics that it is sometimes taught in high school. Anyone who has ever struggled with the "new math" can thank Galois.)

After Galois, group theory was developed into a mature branch of mathematics by the Norwegian mathematician Sophus Lie in the late 1800s. Lie completed the monumental task of cataloging all possible groups of a certain type (now called Lie groups in his honor). With the development of Lie groups, based entirely on abstract mathematical constructs, mathematicians thought they had finally discovered a branch of knowledge that had no practical use whatsoever for the physicists. (Apparently, some mathematicians delight in producing mathematics so pure that it has no practical application.)

They were wrong.

A century later, this "useless" theory of Lie groups would provide the foundation for all of the physical universe!

Lie Groups—The Language of Symmetry

One of Lie's great accomplishments was to catalog all the groups of a certain type into seven varieties.[1] One class of Lie groups, for example, is called O(N).

A beach ball is the simplest example of an object with O(N) symmetry. No matter at what angle the beach ball is rotated, the ball rotates back into itself. We say that this ball has O(3) symmetry (O stands for "othogonal," and 3 stands for the three dimensions of space).

Another example of O(3) symmetry is the atom itself. Because the Schrödinger equation, which is the basis of all quantum mechanics, is invariant under rotations, the solutions to the equation (which are atoms) should also have this symmetry. The fact that atoms have this rotational symmetry is a direct consequence of the O(3) symmetry of the Schrödinger equation.

Lie also discovered a set of symmetries called SU(N), which rotate complex numbers. The simplest example is U(1), which is the symmetry underlying Maxwell's equations (the "1" stands for the fact that there is only one photon). The next simplest is SU(2), which can rotate the proton and the neutron. Heisenberg was the first to show, in 1932, that the Schrödinger equation for these particles, which are very similar except for their charges, can be written so that shuffling these two particles leaves the equation invariant. Another example is the Weinberg-Salam theory, which remains the same if we rotate the electron and the neutrino into each other. Because it rotates two such particles, it has the symmetry group SU(2). Because it also contains the U(1) symmetry of Maxwell, the complete symmetry of Weinberg and Salam is therefore the product $SU(2) \times U(1)$.

Sakata and his collaborators then showed that the strong interactions can be represented by the symmetry group SU(3), which rotates the three subnuclear particles that make up the strongly interacting particles. SU(5), moreover, is the smallest GUT theory that can be written that can shuffle five particles (the electron, neutrino, and three quarks). Naturally, if we have N quarks, then the symmetry group would be SU(N), where N can be as high as is desired.

Perhaps the strangest class of Lie groups, however, are the E(N) groups. It is difficult to imagine a simple example of E(N) symmetry, because these mysterious groups cannot be expressed in terms of ordinary objects. There is no snowflake or crystal that possesses E(N) symmetry. These symmetries were found by Lie strictly through abstract algebraic manipulations having nothing to do with physical objects. The weird feature of these groups is that, for purely mathematical reasons, the highest value N can take is 8. (Explaining why the maximum number is 8 requires an understanding of advanced mathematics.)

This E(8) group is one of the symmetries of the superstring. Because 8 is the maximum number that can be constructed, a strange form of "numerology" is emerging, which is intimately linked to the twenty-six dimensions found in the string model and the ten dimensions found in the superstring. (The origin of this "numerology" is unknown even to the mathematicians. If we could understand why the numbers 8, 10, and 26 continually crop up in the superstring theory, perhaps we could understand why the universe is in four dimensions.)

The key, therefore, to the unified field theory is to adopt the Lie groups as the mathematical framework for unification. Today, of course, it seems easy. Physicists pride themselves on the development of Lie groups and unified field theories, which have astonishing elegance and beauty. However, this was not always the case. Time and again, the majority of physicists have shown themselves to be stubborn, almost pigheaded, resisting vehemently the introduction into physics of larger Lie groups and unification. Perhaps that is one of the reasons why only a handful of physicists have been able to see farther than the rest.

Hostility Toward Unification

In 1941, forty-two years before the discovery of the W-particle and the crowning experimental confirmation of the electro-weak theory, Julian Schwinger of Harvard University mentioned to J. Robert Oppenheimer that the weak and electromagnetic forces could be united into one theory. Schwinger recalls, "I mentioned this to Oppenheimer, and he took it very coldly. After all, it was an outrageous speculation.''[2]

Discouraged, Schwinger nevertheless kept plugging away at this highly mathematical theory. A former prodigy, Schwinger was no stranger to advanced mathematics. He entered the City College of New York when he was only fourteen, transferred to Columbia and was graduated when he was seventeen, and received his Ph.D. at the ripe age of twenty. At age twenty-eight, he became one of the youngest full professors ever at Harvard University.

In 1956, Schwinger showed a remarkably complete version of the electro-weak theory to Nobel laureate Isidor Isaac Rabi of Columbia University. Rabi replied bluntly, ''Everybody hates that paper.''[3] When Schwinger realized that his electro-weak theory violated some experimental data, he threw up his hands and handed over his ridiculous theory to his graduate student Sheldon Glashow. (It was the experimental data that Schwinger saw at that time, and not his theory, of course, that was wrong. Abdus Salam, who won the Nobel Prize with Glashow and Weinberg for the electro-weak theory, later remarked, ''If those experiments hadn't been wrong, he might have gotten the entire thing then and there.''[4])

Although Glashow and his collaborators earned the derision of other physicists, they were on the right track. They united the electron and the neutrino mathematically using SU(2). The electromagnetic theory possessed U(1) by itself, so the complete theory possessed the symmetry $SU(2) \times U(1)$. But almost the entire physics community ignored this theory for several decades.

The same icy reaction greeted the work of Sakata and his collaborators. In the 1950s, years before Gell-Mann introduced the quarks, Sakata and his collaborators went against prevailing opinion and predicted that a sublayer that obeyed SU(3) symmetry existed beneath the hadrons. But Sakata's subnuclear theories were too far ahead of their time to be fully digested by other physicists. His ideas were considered outlandish.

Not unlike some professionals in other fields, when physicists have been laboring over a problem for years, they sometimes tend to be skeptical or even jealous of anyone who suddenly proposes the answer to the entire problem. It's like a detective trying to solve a murder mystery. Imagine someone who has spent months assembling the clues to the mystery. There are many gaps in the evidence, and some of the evidence even seems contradictory. (Moreover, this person is bright, but not a genius.) While he's puzzling over one set of clues, a brash young detective barges into the room, takes one look at the clues, spots a pattern, and blurts, "I know who the killer is!" The plodding detective probably feels a certain amount of resentment, tinged with envy.

After all, the seasoned detective tells the young detective, it is premature to guess the answer when there are so many gaps in the evidence. Anyone, he might say, can propose theories about who is the killer. In fact, he can propose hundreds of reasons why this young detective doesn't appreciate the finer points of being a careful, seasoned detective who doesn't jump to conclusions. His arguments may even convince the young detective, as Oppenheimer did Schwinger, that it's silly to propose that a particular person is the killer.

But what if the young detective is right?

This peculiar hostility comes from the unconscious tendency of most physicists who suffer from the mechanistic process of thinking, often found among physicists in the West, which tries to understand the inner workings of an object by examining the mechanical motions of its individual parts. Although this thinking has produced undeniable success in isolating the laws of particular domains, it blinds one from seeing the overall picture and noticing larger patterns. For decades this mechanistic thinking prejudiced physicists against thinking in terms of unification, which Einstein had been trying to do since the 1920s.

Yang-Mills Theory

Chen Ning Yang, a physicist at the Brookhaven National Laboratory on Long Island in the 1950s, and his colleague, Robert Mills, know all about a good proposal not receiving just attention.[5] For years their proposal demonstrating the power of symmetry and unification was virtually ignored.

Yang was born in 1922 in Hefei, China, where his father was a professor of mathematics. Yang was graduated from Kunming and

Tsinghua universities, but he did not make the pilgrimage to Germany, as did Oppenheimer before him. To the next generation of physicists, it was clear that postwar physics would be dominated by transplanted Europeans, which meant journeying to the United States.

Yang arrived in the United States in 1945 and soon adopted the nickname "Frank," after one of his heroes, Benjamin Franklin. He received his Ph.D. in 1948 at the University of Chicago, a mecca after the war for physics research due to the presence of the Italian physicist Enrico Fermi (who was the first to show in 1942 that a nuclear chain reaction could be controlled, which led to the development of the atomic bomb and nuclear power plants).

As early as 1947, when Yang was a graduate student, he began working on a theory that was more sophisticated and general than Maxwell's theory. In hindsight, it was clear that Maxwell's theory possessed, in addition to invariance under the space-time rotations of relativity discovered by Einstein, another kind of symmetry called U(1). Could this be generalized to SU(2) and higher?

Heisenberg had shown earlier that SU(2) was the symmetry generated by shuffling protons and neutrons in the Schrödinger equation. Heisenberg had created a theory in which the basic equations were "invariant" (remained the same) when protons were turned into neutrons and vice versa. Back then, Heisenberg shuffled these particles through an angle that didn't change whether the proton and the neutron were sitting on the moon or on the earth. This symmetry was insensitive to where the proton and the neutron actually were placed.

Yang, however, asked himself the question: What would happen if we created a more sophisticated theory that is invariant when the proton and the neutron were shuffled through an angle that is different, say, on the moon than on the earth? In fact, what would happen if we performed a different angle of shuffling at each point in space?

This idea—that a different rotation should occur at each point in space—was incorporated into the Yang-Mills theory (also called gauge theory). When Yang and his collaborators worked out the details of the theory in 1954, they found that this local symmetry could be satisfied if they postulated a new mesonlike particle, much like the W-particle of weak interactions.

The reaction of the physics community to their paper, which would become one of the most important papers of the century, was predictable: indifference.

The problem with the Yang-Mills particle, as it was called, was that it possessed too much symmetry. It did not resemble any known particle in nature. For example, the theory predicted that these Yang-Mills particles were perfectly massless, but the conjectured W-meson had a finite mass. Because the Yang-Mills particle did not fit any of the particles found in nature, the theory became a scientific curiosity for the next two decades. In order to make the Yang-Mills theory realistic, physicists somehow would have to break these symmetries while still retaining all the good features of the theory.

As a consequence, for almost twenty years the Yang-Mills theory languished, periodically explored by curious physicists but then discarded again. The theory had no practical application because (a) it was probably not renormalizable (but no one could prove this) and (b) it described only massless particles, while the W-particle had mass. The history of science has many twists and turns, but the neglect of the Yang-Mills theory ranks as one of the great missed opportunities.

Some progress was made when Scottish physicist Peter Higgs noticed that it was possible to break some of the symmetries of the Yang-Mills theory and thereby obtain particles with masses. It now sounded very much like a W-particle theory, but nobody believed that the theory was renormalizable. All this changed with the work of a twenty-four-year-old physicist from Holland.

Gauge Revolution

In 1971 Gerard 't Hooft showed that the Yang-Mills theory, broken according to the method found by Higgs, was renormalizable, which made it a suitable theory of weak interactions. It is no exaggeration to say that the proof that these gauge theories were renormalizable set off a volcanic explosion in the world of physics. For the first time since Maxwell in the 1860s, a theory was created that could unite some of the fundamental forces in nature.

At first, the theory was used with SU(2) × U(1) to describe the electro-weak force. Then it was used in an SU(3) gluon theory to bind the quarks together. Then finally it was used in SU(5) or a higher group to assemble all the known particles into one family.

Physicists, looking back at the "gauge revolution," were stunned to realize that the universe was a lot simpler than they had expected. As Steven Weinberg once remarked:

> . . . although the symmetries are hidden from us, we can sense that they are latent in nature, governing everything about us. That's the most exciting idea I know: that nature is much simpler than it looks. Nothing makes me more hopeful that our generation of human beings may actually hold the key to the universe in our hands—that perhaps in our lifetimes we may be able to tell why all of what we see in this immense universe of galaxies and particles is logically inevitable.[6]

From GUTs to Strings

The GUT model was exciting because it could unite hundreds of particles by postulating the existence of just a few constituent particles consisting of quarks, leptons (electrons and neutrinos), and Yang-Mills particles.

However, problems cropped up. As time went by, atom smashers discovered more and more "fundamental" quarks and leptons, including a fourth quark in 1974. Once again, it seemed that history was repeating itself.

Back in the 1950s, physicists were drowning in an ocean of subatomic particles found in the strong interactions. This led to the discovery of SU(3) and the quark model. In the late 1970s and early 1980s, more quarks were being discovered but, as we saw in chapter 5, they were carbon copies of the previous set of quarks. To physicists, the existence of carbon-copy quarks means that the GUT theory cannot be a fundamental theory of the universe.

Unlike GUT, the superstring theory solves the problem of the proliferating quarks by postulating a single entity—the string—as the fundamental unit of matter with a symmetry $E(8) \times E(8)$.

Why Three Redundant Families of GUT Particles?

Electron Family	**Muon Family**	**Tau Family**
electron	muon	tau
neutrino	muon neutrino	tau neutrino
up quark	strange quark	top quark
down quark	charmed quark	bottom quark

One of the great embarrassments of the GUT theory is that it fails to explain why there are three identical families of particles. In the superstring theory, however, these redundant families can be explained as different vibrations of the same string.

(Lie found that, in addition to the SU(N) groups, there was another class of groups that he called E(6), E(7), and E(8) (E stands for "exceptional"). These groups were exceptional because, instead of going on forever, they simply stopped at E(8). This group, which contains the GUT symmetry, is important for strings.)

Origami and Symmetry

The superstring theory works so well because it has two sets of powerful symmetries, conformal symmetry and supersymmetry. Origami can be used to illustrate the first symmetry. (We will discuss the second symmetry in the next chapter.)

Earlier, we saw that Tinker Toys are useful in calculating the S-matrix for point particles. With sticks and a joint, we can create an infinite number of Feynman diagrams that, when summed up, yield the S-matrix. However, there is no rhyme or reason behind most of these Feynman diagrams. We simply blindly attached Tinker Toys in all possible ways. Fortunately, for simple theories, such as QED, it takes only a few diagrams to get spectacular agreement with the data.

In a quantum theory of gravity, however, it takes tens of thousands of these diagrams to represent even one loop diagram. And most of these diagrams are divergent. Can nature really be this complicated? Anyone who has spent years toiling over these diagrams, churning out thousands of pages of dense equations, feels that there must be an underlying pattern.

The superstring theory provides this symmetry, allowing these thousands of diagrams to collapse into just a few. The enormous advantage of these diagrams is that they can be stretched and shrunk, like rubber, without changing their value. For example, at the first loop level, instead of having tens of thousands of Feynman diagrams, we only have one. All the tens of thousands of different one-loop Feynman diagrams can be shown to be equal to one another by stretching them.

Clearly, this symmetry provides an enormous simplification of the theory. This symmetry, in fact, is so powerful that it cancels thousands of divergences, resulting in a finite S-matrix.

Broken Symmetry

If nature were symmetrical, then the work of physicists would be much easier. The unified theory would be obvious, because there would be only one force, not four. However, nature is full of surprises in the form of broken symmetries. For example, the natural world is not perfectly crystalline or uniform but is filled with irregular galaxies, lopsided planetary orbits, and the like. The world is full of examples of where symmetry is hidden because it is broken. (In fact, the universe would be a rather dull place if symmetry was never broken. Humans could not exist [because there would be no atoms], life would not be possible, and chemistry itself would collapse. It is symmetry's breaking, therefore, that makes the universe so interesting.)

The study of broken symmetries explains, for example, the freezing of water. Water in liquid form possesses great symmetry. No matter how we rotate water, it remains water. In fact, even the equations governing water have the same symmetry. However, as we cool water slowly, random ice crystals form in all directions, creating a chaotic network that eventually becomes solid ice. Here is the essence of the problem: Although the original equations possessed great symmetry, the solutions to the equations do not necessarily have this symmetry.

The reason why these quantum transitions take place is that nature always "prefers" to be in the lowest energy state. We see evidence of this all the time; for example, water flows downhill because it is trying to reach a lower energy state. Quantum transitions occur because the system started originally in the wrong energy state (sometimes called the "false vacuum") and would prefer to make a transition to a lower energy state.

Symmetry Restoration

At this point, analyzing the broken pieces of a symmetry to reveal the hidden symmetry may seem a hopeless task. However, there is one way in which we can recover the original symmetry: by heating the substance. By heating ice, for example, we recover water and O(3) symmetry. Likewise, if we want to restore the hidden symmetries of the four forces, we must reheat the theory—go back to the Big Bang, where temperatures were high enough to restore the broken symmetry of the superstring. Of course, we cannot physically reheat the universe and re-create the conditions of the Big Bang. However, by studying the Big Bang, we can analyze an era when the universe had its symmetries intact.

In fact, physicists suspect that at the beginning of time, temperatures were so hot that all four forces were merged into one. As the universe cooled, however, one by one the symmetry holding together the four forces began to break down.

In other words, the reason why we see four forces today is that the universe is so old and so cold. If we had witnessed the Big Bang itself, then, and if the theory is correct, we would have seen all matter manifesting the symmetries of the superstring, such as supersymmetry, which we will explain in the next chapter.

Yet, if physicists claim that supersymmetry is the key, and if supersymmetry is such a simple theory, why did it elude physicists for so many years?

8

Supersymmetry

THE MAN who has figured most prominently in the discovery of superstrings is John Schwarz of the California Institute of Technology.

Like some other leading superstring physicists, John Schwarz comes from a family of scientists. His father was an industrial chemist and his mother was a physicist at the University of Vienna. His mother even secured a job with Madame Curie in Paris, but the great chemist died before she could begin. John's parents were from Hungary, but, with the massive anti-Jewish sentiment rising under the Nazis in Europe, they fled Europe in 1940 and settled in the United States. John was born in 1941 in North Adams, Massachusetts.

He began his undergraduate work at Harvard as a math major but was graduated as a physics major in 1962. "I was beginning to get frustrated with mathematics," he recalls. "While it was a lot of fun, I really didn't see the point in it all. But trying to answer the questions posed by nature, that seemed to me to be more focused, and also more satisfying."[1]

After Harvard, he went to graduate school at the University of California at Berkeley. "That was the hotbed of theoretical physics in those days," he remembers fondly. The S-matrix theory was at its height, and both he and David Gross of Princeton worked under

Geoffrey Chew. Among the future luminaries at Berkeley at that time were junior faculty members Steven Weinberg and Sheldon Glashow. "When Weinberg entered the room," says Schwarz, "there was a certain aura about him. You knew he was an important person."[2]

Leaving Berkeley with his Ph.D. in 1966, Schwarz went to Princeton University, where he worked with two young French physicists from Paris, Andre Neveu and Joel Scherk. Schwarz, with these two Frenchmen, produced a series of seminal superstring papers. In 1971, Neveu and Schwarz realized that there was a fundamental flaw in the Beta function proposed by Veneziano and Suzuki: Their theory could not describe all the "spinning" particles found in nature.

All objects have "spin," or angular momentum—everything from galaxies (which may take millions of years for one rotation) to subatomic particles (which can rotate millions of times in just one second). Familiar objects such as a spinning top can spin at any rate. A record player, for example, can be adjusted to spin at 33⅓ rpm as easily as 78 rpm with the flip of a dial.

However, in the quantum world, the spin of an electron does not occur in arbitrary amounts. Just like light, which can occur only in discrete bundles called photons, subatomic particles can spin only with certain amounts of angular momentum.

In fact, quantum mechanics divides all the particles of the world into just two types: *bosons* and *fermions.*

As an example of fermions, look at your body. The electrons and protons that make up the atoms in your body are all fermions. Everything you see around you, including the walls and sky, is made of fermions, which have half integral spin: $1/2$, $3/2$, $5/2$, and so on, measured in units of Planck's constant. Fermions are named in honor of Enrico Fermi.

As an example of bosons, think of the gravity that prevents you from being spun into outer space. Or think of light itself. Without bosons, the universe would be dark and without any gravity to hold the stars together. Bosons have integral spin: 0, 1, 2, and so on. Bosons are named after Indian physicist Satyendra Bose.

Fermion	Spin:	Boson	Spin:
electron	½	photon	1
neutron	½	graviton	2
proton	½	W-particle	1
neutrino	½	pi meson	0
quark	½		

The spin of a particle is quantized and measured in units of Planck's constant divided by 2π, an exceedingly small number. For example, the electron has spin $\frac{1}{2} \times \frac{h}{2\pi}$, while the photon has spin $1 \times \frac{h}{2\pi}$.

Today, we realize that Nambu's string theory, which explained the origin of the Veneziano-Suzuki Beta function, was only a bosonic string. Neveu, Schwarz, and Ramond completed the theory by inventing a *fermionic* string to accompany the bosonic one. The Neveu-Schwarz-Ramond theory (with a slight modification) became the superstring theory of today.[3]

This theory of Neveu, Schwarz, and Ramond predicted a new S-matrix that had even better properties than the old S-matrix of Veneziano and Suzuki, but the origin of these near-miraculous properties was obscure. Whenever there are such marvelous "coincidences," physicists suspect that a hidden symmetry is responsible for it.

In 1971, Bunji Sakita of the City College of New York and Jean-Loup Gervais of the École Normale in Paris found a partial answer to this puzzle. They showed that the Neveu-Schwarz-Ramond theory indeed possessed a hidden symmetry that was responsible for its astonishing properties. These pioneering discoveries marked the beginning of supersymmetry. (Supersymmetry was proposed simultaneously by two Soviet physicists, Yu. A. Gol'fand and E.P. Likhtman, although their work was not appreciated in the West at that time.)

The supersymmetry Gervais and Sakita discovered was the most unusual symmetry ever found. For the first time, a symmetry was created that could rotate a bosonic object into a fermionic object. Eventually, this meant that all bosonic particles in the universe had a fermionic partner. (Their symmetry, however, was not yet complete because it was only a two-dimensional symmetry. The theory was in two dimensions because when a one-dimensional string moves, it sweeps onto a two-dimensional surface, a strip.)

Tremendous excitement was generated by this new superstring theory and the discovery of an entirely new symmetry that interchanged fermions with bosons. However, in the mid-1970s the theory took a nosedive.

The Severest Critic

As mentioned previously, the discovery that the bosonic string of Nambu existed only in twenty-six dimensions and that the superstring of Neveu-Schwarz-Ramond existed only in ten dimensions killed the model in the mid-1970s. Schwarz and his collaborator, Michael Green, seemed to be the only ones promoting string research. Nobody, it seemed, wanted to do research in ten-dimensional space-time.

Schwarz was convinced, however, that the difficulties could be ironed out. He remembers a conversation he had with Richard Feynman during those lean years in which Feynman said that whenever we propose any theory, we must be our own severest critic. Undoubtedly, says Schwarz, Feynman said this as a favor, to discourage him from wasting his productive years on the string theory, which was likely a dead end. To Schwarz, however, this had the opposite effect. "Feynman didn't realize this, but in my string work, I tried to be very critical, but I couldn't find anything wrong with it!"[4]

The development of the theory suffered another setback during those years with the unexpected death of Joel Scherk. Michio remembers first meeting Scherk in 1970, when Scherk had just left Princeton and was visiting Berkeley. They worked together and published the first paper on the singularity structure of multiloop diagrams.[5] Scherk was an unconventional but gentle soul who seemed perfectly at home with the antiwar counterculture then flourishing at Haight-Ashbury in San Francisco and Telegraph Avenue in Berkeley. After leaving Berkeley, he returned to France, typically in a most unconventional way. First he journeyed to Japan, where he stayed for several weeks in a Buddhist monastery, meditating along with the monks in ascetic fashion. Then he journeyed to France via the Trans-Siberian Railway. It was during this period that he developed a severe case of diabetes. Apparently because of this, as well as mounting personal problems, he committed suicide in 1980.

Rise of Supergravity

Although the string rapidly fell into disfavor, other physicists tried to salvage supersymmetry as a symmetry of ordinary point particles. The symmetry that changed fermions into bosons and vice versa was just too good to pass up.

Inspired by the work of Gervais and Sakita, in 1974 Bruno Zumino (now at Berkeley) and Julius Wess (of Karlsruhe University in West Germany) showed how this new symmetry could be extracted from the string and reduced to a simple point particle theory defined in four dimensions (a conventional quantum field theory). They took one of the simplest field theories—a spin 0 boson interacting with a spin ½ fermion—and showed that it could be made supersymmetric. More important, they showed, simply and cleanly, that supersymmetry killed many unwanted divergences found in quantum field theories of point particles. Just as the SU(N) symmetries of the Yang-Mills theory killed all of the divergences of the W-particle theory, supersymmetry killed many (but not all) of the divergences of point particle theory.

Imagine the Feynman diagram on the left side of the figure, which diverges because it has a fermion circulating in the interior loop. Wess and Zumino discovered, to their surprise, that this divergence could be made to cancel the divergence of the diagram on the right, which has a boson circulating in the inner loop. In other words, the divergence of the left loop cancels beautifully the divergence of the right loop, leaving a finite result. Here we see the power of symmetry in canceling a divergence.

Similarly, symmetry can be used to solve problems outside the realm of physics. Let's say, for example, that a seamstress has sewn a wedding dress. However, just before the wedding, the seamstress discovers that the dress is slightly lopsided. She has two options. She can retrieve all the patterns, tediously compare the lopsided pieces

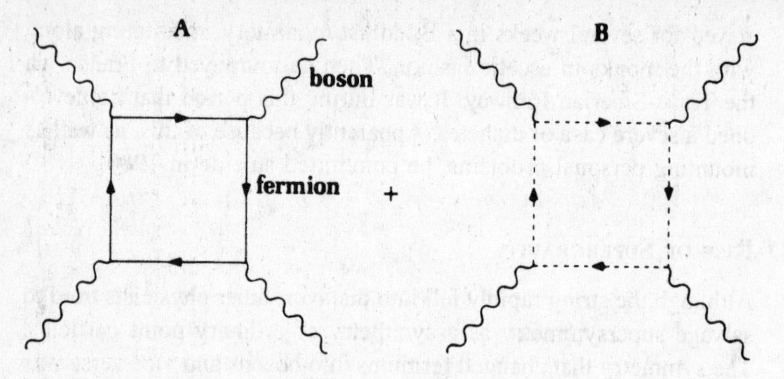

In diagram A, the interior solid line represents a fermion. The divergence of diagram A cancels the divergence of diagram B, which contains a boson (represented by a wavy line). Thus, the sum of the two diagrams is finite.

with the originals, and carefully cut off the excess. Or she can use the power of symmetry and simply fold the wedding dress in half, making the two sides match, and cut off the excess. Symmetry can be used to cancel the divergences of left and right halves, and isolate the unwanted excess.

Similarly, supersymmetry allows us to match the two sides of divergent Feynman diagrams until they cancel each other perfectly, with no excess.

Because supersymmetry was easy to adapt to point particle theories, in 1976 three physicists working at the State University of New York at Stony Brook tinkered with Einstein's old theory of gravity. Building on the success of Wess and Zumino, they successfully added a fermionic partner to the graviton and created a new theory they christened "supergravity."

Supergravity, although just a small part of the superstring (it emerges when we take the length of the string to be zero, that is, a point), is interesting in its own right. In a sense, it represents a halfway station between Einstein's theory of gravity and the superstring.

Because gravity has two units of spin, it must have a partner with half-integer spin, 3/2, which physicists called the "gravitino" (small gravity).

Supergravity created quite a stir when it was first proposed, since it was the simplest nontrivial extension of Einstein's equations in sixty years. In fact, one magazine published a whimsical picture of two of its creators, Peter van Nieuwenhuizen and Dan Freedman, leaping in the air, as if they were defying the laws of gravity and creating some form of antigravity.[6]

Although supergravity initially created great expectations, the theory showed unmistakable problems in uniting the forces of nature. For one thing, the theory was simply too small to accommodate all known particles. The smallest Lie group that can accommodate all known particles is SU(5). However, the largest Lie group that could be fit into supergravity is O(8), which is too small to include all the quarks and leptons in a true GUT theory. The largest supergravity cannot accommodate both quarks and leptons simultaneously.

In summary, even though the supergravity theory was appealing, its symmetry was simply too small to eliminate the divergences or to incorporate both quarks and leptons.

Princeton String Quartet

In the late 1970s, physicists realized that supergravity was a small piece of the superstring theory. For example, the supergravity theory emerges from the superstring theory if we use the smallest closed superstring. However, the superstring theory was considered too mathematical to be realistic.

It took the discovery in 1984 by Green and Schwarz that the theory is free of all anomalies to set off the explosion of interest in superstrings. Very rapidly, the superstring theory, which was considered dead by the vast majority of physicists worldwide, sprang back to life as the most powerful quantum field theory ever constructed.

By then it was becoming clear that an enormous symmetry group would be necessary to eliminate all the divergences in gravity, and the superstring theory had the largest set of symmetries that physicists had ever seen.

Four physicists at Princeton—David Gross, Jeffrey Harvey, Emil Martinec, and Ryan Rohm—discovered a new superstring with the symmetry group E(8) × E(8), which had even better properties than the Green-Schwarz superstring. The Princeton group (dubbed the

"Princeton string quartet") showed that the E(8) × E(8) string was compatible with all the earlier GUT theories and hence was consistent with all known experiments. E(8) alone is much larger than SU(5), and therefore the theory not only gobbles up all known GUT-type theories, it predicts thousands of new particles that have never been seen. The Princeton superstring currently is the leading candidate for a theory of the universe.[7]

Supernumbers

The superstring theory is probably the craziest theory ever proposed, and its underlying symmetry, called supersymmetry, is equally strange.

Ironically, supersymmetry has never been found in nature. So far it exists only on paper, but it is so beautiful and compelling a theoretical tool that most physicists take it for granted that supersymmetry eventually will be discovered.

But if supersymmetry is such a beautiful symmetry, why wasn't it discovered years ago? There is a simple but deep reason for this, which goes back to the origin of human society and how we count with our fingers.

Since humans began to count thousands of years ago, we have assumed that numbers correspond to tangible, real things. We know that numbers can be added; we are taught that 5 sheep added to 2 sheep will produce 7 sheep. As society grew increasingly complex, rules had to be invented to add and subtract larger and larger numbers. The Romans required sophisticated methods of adding and dividing in order to collect taxes and trade with other lands. In this way, the earliest rules of arithmetic were developed as a way to count goods that could be traded or sold.

The ancients found that numbers can be added or multiplied in any order. For example, we know that $2 \times 3 = 3 \times 2 = 6$. We know these relations are true because we can count objects with our fingers and demonstrate that they are correct. But why should the generalization of this relationship between numbers hold up for the unified field theory?

One reason why supersymmetry was not discovered for so many years was that we must create a new set of numbers that do not obey these "commonsense" rules. In particular, let's say we want to invent a new number system, called Grassmann numbers, in which $a \times b = -b \times a$. The minus sign, although innocent enough, has far-reaching implications when applied to theoretical physics.

This means, for example, that $a \times a = -a \times a$. At this point, you may object, because this means that $a \times a = 0$. Normally, one would say that this means that $a = 0$. However, for Grassmann numbers, this is not so.

Thus, a meaningful system of "arithmetic" can be constructed in which $a \times b = -b \times a$. The system can be shown to be mathematically self-consistent and a completely satisfactory system of arithmetic. This bizarre system of numbers requires us to expand upon the last ten thousand years of arithmetic.

Supersymmetry, like all other developments in the history of the unified field theory, creates a special unification of its own: It unites the concept of a real number with a Grassmann number and yields a "supernumber."

In summary, supersymmetry was not discovered earlier partly due to a basic, unconscious prejudice among physicists against using Grassmann numbers to explore nature. In fact, Sophus Lie, the great Norwegian mathematician who thought he had cataloged all possible types of groups, missed the supersymmetric groups, which are based on Grassmann numbers.

Of course, one might be tempted to protest that these abstract constructions seem devoid of physical content. However, Grassmann numbers are eminently practical. Because Grassmann numbers describe fermions, the human body is made up of particles that can be described only by Grassmann numbers.

Supersymmetry at the Beginning of Time

Unfortunately, there is no experimental evidence for supersymmetry's existence. If supersymmetry existed as a physical symmetry at our energy scale, then the electron, with spin ½, would have a partner, a spin 0 meson. However, this is not verified experimentally. Not surprisingly, supersymmetry often has been called "a solution looking for a problem," because, despite its beauty and elegance, nature seems to ignore it within the energy range of our machines.

The advocates of supersymmetry, however, are not fazed. If supersymmetry is not yet found at low energies, then obviously, they reason, we must build larger atom smashers and probe deeper into the interior of the proton. The problem, they say, is not the absence of supersymmetry but rather the lack of powerful enough machines that can probe larger and larger energy scales.

To discover supersymmetry, as well as other mysteries of the subatomic world, the United States government had planned to build the largest machine in the history of pure science—the superconducting super collider (SSC)—but the project was canceled by Congress in 1993. Yet, given its vast scale, the SSC still merits discussion.

Antaeus

Although the SSC was an awesome project—one that had been compared to the building of the pyramids—particle physics had humble origins.

In the 1920s, physicists studied elementary particle physics by examining cosmic rays (radiation emanating from outer space, the origin of which is not yet well understood) using equipment costing a fraction of a percent of the present cost of atom smashers.

Historically, cosmic ray experiments were conducted by sending up photographic plates in large balloons. It was a tedious process to send balloons into the upper atmosphere, reclaim them, develop the film, and then spend months examining the emulsion for possible tracks left by high-energy cosmic rays. It was a slow grab-bag approach to experimental physics, because physicists never knew ahead of time what they would find. (The pi meson of Yukawa, for example, which mediates the strong force, was first discovered by looking at the tracks left by cosmic rays, but only after months of careful work.)

It was a nuisance, moreover, to analyze the tracks of random cosmic rays. The energies of these cosmic rays could not be predicted, and it was impossible to perform controlled experiments with cosmic rays of unpredictable energy.

All this changed in the 1930s with the invention of the first atom smasher, the cyclotron, by Ernest Lawrence of the University of California at Berkeley. This machine was only a few inches across and produced a feeble beam of energy, but it could create made-to-order beams similar to cosmic rays right in the laboratory.

This evolution can be compared with our human evolution, where we spent hundreds of thousands of years foraging for food in the forests. Our early ancestors never knew ahead of time what kind of fruits or game they might find. It was a painful, random process. The great revolution, of course, occurred when we learned the laws of agriculture and began to harvest grain and tend sheep and cattle, thereby securing food sources under controlled conditions rather than being at the mercy of fate.

Throughout the 1980s, the Department of Energy considered proposals for the SSC, which likely would have cost more than $11 billion and require a staff of about three thousand scientists and engineers.

The goal was to build a machine that would allow physicists to explore the question of whether the four fundamental forces were originally one. Consequently, the SSC would have been not only the most expensive but the biggest piece of scientific apparatus ever built.

The magnetic coils of the SSC would produce magnetic fields of 6.6 teslas, or about 130,000 times stronger than the earth's magnetic fields. Such powerful magnetic fields can be produced due to a quantum effect called "superconductivity," in which the electrical resistance of metals drops to zero when temperatures drop to near absolute zero. The magnets would be cooled by liquid helium, which is held at 4.35 degrees above absolute zero.

The machine itself would be housed in a narrow circular tunnel about twenty feet wide and perhaps two hundred miles long, which would be placed underground (to absorb the intense radiation created by the machine). Inside this tunnel would be a series of powerful magnets that could bend the path of the particles as they circulated along this ring.

The heart of the SSC would consist of two distinct tubes, no more than two feet in diameter, running throughout the length of the tunnel. Within these two tubes, two beams of protons would travel in opposite directions and be accelerated to enormous energies by electrodes located along the beam's path. (The beams would be accelerated within fifteen minutes of startup and travel 3 million times around the tube, attaining velocities within a fraction of the ultimate velocity: the speed of light.)

These two beams of protons would circulate in opposite directions until electromagnetic gates were opened and the two beams collided head on, creating intense temperatures and conditions not seen since the Big Bang. (For example, the impact would create energies of 40 trillion electron volts.)

European nations, themselves not large enough to build such projects, banded to form CERN near Geneva, but the SSC would have been sixty times larger than the biggest machine at CERN.

Scientists had hoped to test a host of new theoretical ideas with the SSC. The old electro-weak theory of Weinberg and Glashow would be the easiest to test. In the long run, however, scientists had hoped to discover clues that would help us understand GUT theories and possibly superstrings. Because both GUT and superstring theories attain unification at energies a quadrillion times greater than would have been found at the SSC, we could only hope to see glimmers of these two theories.

Although with the SSC we were fast approaching the practical limit to which nations on this planet could afford to delve into the realm of subnuclear physics, other avenues are opening up all the time.

For example, the United States is now launching orbiting laboratories on satellites that can peer into faraway galaxies in search of black holes and the remnants of the Big Bang. In fact, we may have to use echoes of creation itself as the "laboratory" in which to collect our data.

This process of linking the experimental data to the theory is of paramount importance to any theory, especially to one that claims to unite all known forces. As physicist Maurice Goldhaber remarked, borrowing from Greek mythology, "Antaeus was the strongest person alive, invincible as long as he was in contact with his mother, the earth. Once he lost contact with the earth, he grew weak and was vanquished. Theories in physics are like that. They have to touch ground for their strength."[8]

Answering the Critics

As Julian Schwinger once said, commenting on the GUT theory, "Unification is the ultimate goal of science, it's true. But that it should be unification *now*—surely that's the original definition of hubris. There are a heck of a lot of energies we haven't gotten to yet."[9]

Although Schwinger meant this as a criticism of the GUT theory, it could apply equally well to the superstring theory. Although the theory now offers the only hope of providing a comprehensive framework to describe the laws of the universe, some critics of superstrings point out that perhaps even the SSC would not have been large enough to test fully the consequences of physics at the Planck scale of 10^{19} billion electron volts.

Schwarz, however, remains undaunted. "Certainly, this is not just a theory of physics at 10^{19} billion electron volts, but if it is correct, it is a theory about physics at every scale. We need to develop our mathematical tools to get out the low energy consequences."[10]

In other words, the problem is not so much our inability to build large machines, but that our mathematical understanding of how a ten-dimensional universe became a four-dimensional universe is too primitive at this point. Our next step, then, is to investigate supersymmetry by studying the greatest "laboratory" of all: the universe at the beginning of time.

III

Beyond the Fourth Dimension

9

Before the Big Bang

EVERY SOCIETY has its mythology about the origin of time. Many of these myths refer to the fiery origin of the universe, when the gods waged war in the heavens over the destiny of the newly created earth. The ancient Norse myths concerning the origin and death of the universe are filled with colossal battles between giants, gods, and trolls, leading up to the epic Ragnarok, the death of the gods themselves.

Now, for the first time, scientists are able to make sensible statements about the creation based on physics rather than on mythology. What has been especially exciting about cosmology (the study of the origin and structure of the universe) is the interplay between quantum mechanics and relativity, which has opened surprising new vistas of which Einstein never dreamed.

But perhaps the most spectacular conclusion of the superstring theory is that it can actually make statements about what happened before the Big Bang, at the beginning of time. The superstring theory, in fact, views the Big Bang as a by-product of a much more violent explosion, the breakdown of a ten-dimensional universe into a four-dimensional one.

The Big Bang

The origin of the Big Bang theory can be traced to a mistake that Einstein made in 1917, which he later called "the biggest blunder" in his life.

In 1917, two years after he wrote down his celebrated general theory of relativity, Einstein found a most disturbing result. Every time he solved his own equations, he found that the universe was expanding. In that era, it was "common knowledge" that the universe was eternal and static. Even the idea that galaxies could exist beyond our own Milky Way was considered heresy verging on science fiction. Much to his chagrin, Einstein found that his equations flew in the face of common sense. Were his equations wrong?

He found the idea of an expanding universe so alien that he was forced to conclude that his equations were incomplete. Einstein then added a "fudge" factor into his equations to balance the tendency of the universe to expand. Even Einstein, the great revolutionary who overturned three hundred years of Newtonian physics, couldn't get himself to believe in his own equations and had to "cheat."

In 1922, Soviet physicist Alexander Friedman found perhaps the simplest solution to Einstein's equations that gives us the most elegant description of the expanding universe. However, like Einstein's solutions, no one took his ideas seriously because they ran counter to the conventional wisdom of the time.

Finally, in 1929, the American astronomer Edwin Hubble, after years of work with the one-hundred-inch Mount Wilson telescope, announced his dramatic findings: Not only were there millions of galaxies in space far beyond our Milky Way, but they were all rushing away from the earth at fantastic speeds. Einstein and Friedman had been right all along!

Two years later, in 1931, Einstein dropped this "fudge" factor and reintroduced his old theory of the expanding universe, which he had abandoned fourteen years earlier.

Hubble found that the farther a galaxy was from the earth, the faster it was traveling away from the earth. To measure the enormous velocities of these galaxies, scientists rely on the Doppler effect. (According to the Doppler effect, light or sound waves from an object coming toward you have a higher frequency than light or sound waves from a receding object. This is what accounts for the fact that the sound from a speeding train drops so dramatically when it whizzes past you.)

Hubble verified that this Doppler effect was happening to the light from distant stars, creating a "red shift" of starlight. (If the stars were coming toward the earth, there would be a "blue shift," which is not seen experimentally.)

The expanding universe often is compared to a balloon being blown up. Imagine that plastic spots have been glued onto the surface of the balloon. As the balloon is inflated, the spots (galaxies) move away from one another. We live on the surface of the balloon, so it appears that all the stars are moving away from us.

The expanding universe also explains a paradox that has puzzled astronomers for years: Why is the night sky dark? In 1826, Heinrich Olbers wrote a paper in which he argued that if there are an infinite number of stars, the light coming from them should fill the night sky. No matter where we look in the night sky, we should be blinded by the brilliant light. But in an expanding universe, energy is lost by the red shift, and stars have a finite lifetime, so we are not blinded by the night sky.

Although this "expanding universe" model has been verified experimentally, Einstein's theory says little, if anything, about how the Big Bang took place and what happened before it. To answer these questions, we must appeal to the theory of GUTs and superstrings.

The GUT Early Universe

For a string theorist, one of the purposes of studying cosmology is to use quantum symmetry breaking as a probe for the early universe. Our universe today is horribly unsymmetrical, with all four forces looking totally dissimilar, but we now know that the reason for this is that our universe is so old.

At the beginning of time, when temperatures were incredibly hot, our universe must have been perfectly symmetrical. All forces were united in one coherent force. As the universe exploded and rapidly cooled, however, the four forces split apart, one by one, until all four forces had no obvious resemblance to one another—as they are today.

This means that we can use the event of the Big Bang as a "laboratory" to test our ideas about how symmetry should be broken. For example, as we go back in time, eventually we will hit temperatures where the GUT symmetry was unbroken. This, in turn, allows us to explain one of the most puzzling secrets of the universe: What happened at its birth?

For example, we know that at the beginning of time, gravity, the electro-weak force, and the strong force were probably all part of a single force.

When the universe was perhaps only 10^{-43} seconds old, and only 10^{-33} centimeters across, matter and energy probably consisted of unbroken superstrings. Quantum gravity, as described by the superstring, was the dominant force in the universe. Unfortunately, no one was there to witness the event, because at that time the universe could fit easily into a proton.

But at the incredible temperature of 10^{32} Kelvin (a thousand trillion trillion times hotter than the temperatures found in our sun), gravity separated from the other GUT forces. Like water droplets condensing from a cloud of vapor, the forces began to separate.

At that point, the universe was doubling in size every 10^{-35} seconds. As it cooled, the GUT force itself began to break, with the strong force peeling off from the electro-weak force. The universe was about the size of a bowling ball, but was expanding rapidly.

When the universe reached a temperature of 10^{15} degrees K, 10^{-9} seconds after creation, the electro-weak force broke into the electromagnetic force and the weak force.

At this temperature, all four forces had separated from one another, and the universe consisted of a "soup" of free quarks and leptons and photons.

A little later, when the universe cooled still further, the quarks combined to form protons and neutrons. The Yang-Mills fields condensed into that sticky "glue" we mentioned earlier that binds quarks into hadrons. Finally, the quarks in this cosmic "soup" condensed into protons and neutrons, which eventually formed nuclei.

Three minutes after the creation, stable nuclei began to form.

Three hundred thousand years after the Big Bang, the first atoms were born. The temperatures dropped to 3,000 K, the point where hydrogen atoms can form without being ripped apart by collisions. At that time, the universe finally became transparent—light could travel for light-years without being absorbed. (Prior to this time, it was not possible to see through space. Light would simply be absorbed, making distant observation with telescopes impossible. Although we think of space as being dark and transparent, previously space actually was opaque, like a dense fog.)

Today, 10 to 20 billion years after the Big Bang, the universe looks horribly unsymmetrical and broken, with all four forces dramatically different from one another. The temperature of the original fireball has now cooled down to 3 degrees K, which is close to absolute zero.

It is possible, therefore, to describe the overall scheme of unification by the way in which forces disentangle themselves from one another, step by step, as the universe cools. Gravity breaks off first, then the strong force, followed by the weak force, leaving only the electromagnetic force unbroken.

Glashow summed up how GUT theorists view the birth and death of the universe when he said: "Matter first appeared 10^{-38} seconds or so after the Big Bang, and will all disappear maybe 10^{40} seconds from now."

Echo of the Big Bang

It might seem eerie that we humans are able to talk so glibly about temperatures and events so cataclysmic that they would rip apart our earth (even our galaxy). In fact, physicist Steven Weinberg, writing about the birth of the universe, admitted frankly, "I cannot deny the feeling of unreality in writing about the first three minutes as if we really know what we are talking about."[1]

Ultimately, these statements about the early universe are still just theories. The fact remains, however, that no matter how fantastic the details of the creation are, experimental evidence is piling up to confirm that such an event took place according to the predictions of the quantum theory and the theory of relativity.

In particular, Russian physicist George Gamow predicted in the 1940s that there might be a way to verify experimentally that the Big

Bang actually took place. Gamow maintained that the original radiation left over from the Big Bang should still be circulating around the universe, although its temperature would be quite low after 10 to 20 billion years. He predicted that this "echo" from the Big Bang would be distributed evenly around the universe, so it would appear the same no matter where we looked. In 1948 his collaborators Ralph Alpher and Robert Herman even calculated the temperature to which the cosmic fireball had now cooled down: 5 degrees Kelvin.

In 1965, there was spectacular verification of the Gamow-Alpher-Herman prediction of this "echo," or background radiation left by the original Big Bang. At the Bell Telephone Laboratories in Holmdel, New Jersey, scientists constructed a huge radio antenna, the Holmdel Horn Antenna, that would relay messages between the earth and communications satellites. Much to their dismay, scientists Arno Penzias and Robert Wilson discovered that the antenna was picking up a bothersome background radiation in the microwave range. No matter where the antenna was pointed, this strange radiation was being received. Annoyed, the scientists checked all their data and cleaned their equipment (they even cleaned the pigeon droppings from the antenna) but this radiation persisted.

Finally, instruments were put aboard high-altitude jet airplanes and balloons to get rid of interference from the earth, but this strange signal became even stronger. When the scientists plotted the relationship between the intensity of the radiation and the frequency, it resembled the curve predicted so many years earlier by Gamow and others. The measured temperature of 3 degrees was remarkably close to the original prediction of the temperature of the cosmic fireball. Penzias and Wilson discovered, much to their delight, that this radiation was precisely the background "echo" that had been predicted. This 3-degree radiation is still the most conclusive evidence that the universe started with a cataclysmic explosion. This brilliant piece of detective work, for which Penzias and Wilson won the Nobel Prize in physics in 1978, was a stunning confirmation of the Big Bang.

Yet another way to study the curious properties of general relativity and the early universe is to examine the warping of space-time caused by massive dead stars: black holes.

Black Holes

What is a star? Quite simply, it is a gigantic atomic furnace that releases the energy stored in the strong force. A star burns hydrogen as a fuel, creating the "ash" of helium. The basic equations for the burning of hydrogen and other elements in the sun and other stars were worked out by Hans Bethe in 1939, for which he received the Nobel Prize in 1967.

A star exists as a stable object due to the delicate balance between its internal nuclear fires, which tend to blow apart the star, and the gravitational force, which tends to crush the star down to a point. In other words, stars exist because there is a balance between the energy created by the strong force, which tends to be explosive, and the gravitational force, which is implosive.

This delicate balance, however, is destroyed when the star's nuclear fuel (basically hydrogen, helium, and the lighter elements) is used up over billions of years. Once the nuclear fuel is exhausted, the gravitational force takes over. If the gravitational force is great enough, the star will collapse, crushing the atoms into a dense ball of neutrons, creating a dead star, called a "neutron star."

A neutron star is so dense that the individual neutrons of the star are actually "touching" one another. A neutron star, therefore, is a solid mass of nucleonic matter, without any atoms or space between the orbiting electrons and the nucleus. To imagine the enormous contraction necessary to produce a neutron star, think of compressing the entire mass of the sun, which is much larger than the earth, down to the size of Manhattan.

Numerous neutron stars have been found by astronomers. Back in 1054, Chinese astronomers observed a gigantic mysterious explosion in the heavens that was visible even in daytime. Today, we now know that it was a rare "supernova," a cataclysmic explosion of a star that generates more energy than an entire galaxy. This supernova occurred in the Crab Nebula, and at the center of this explosion is now a neutron star.

However, if the original star was massive enough (perhaps several times larger than the mass of our own sun), the neutron star itself will be unstable; the gravitational force will be so great that the neutrons will be pushed into one another, finally crushing the neutron star down to the size of an infinitesimal point. This point particle is the black hole.

The viselike grip of the black hole's gravitational field is so great that nuclei are ripped apart and light itself cannot escape but is forced to orbit the star. This means that light from these dead stars cannot be viewed directly, so the holes appear black, hence the name. If we think back to Newton's famous diagram showing the orbit of a rock when thrown from a mountaintop, we must now replace the rock with a beam of captured light orbiting a black hole.

Like the Cheshire cat in Alice's Wonderland, the black hole disappears from view, leaving only its ''smile,'' the distortion of space-time as a result of the intense gravitational force.

Moreover, the severe warping of space-time created by the black hole resembles the early universe. Time, for example, slows down as you near the center of the black hole. This means that, if you were to fall into a black hole, it might appear that you were slowing down until you were frozen in time, taking thousands of years to fall into the center in slow motion. The closer you come to the center, the slower time becomes. In fact, at the center of the black hole, time itself supposedly stops. (In reality, this probably means that general relativity simply breaks down at the center of the black hole, and the superstring theory must take over as we calculate quantum corrections to general relativity.)

Black holes were first postulated theoretically as a consequence of general relativity by J. Robert Oppenheimer and his student Hartland Snyder in 1939. Although even Oppenheimer was taken aback at the astonishing conclusions of general relativity that stretched the limits of the imagination, in the summer of 1994 the Hubble Space Telescope discovered that the galaxy M87 (50 million light-years from the earth) contained a black hole. In January 1995, a second black hole was identified in the galaxy NGC4258 (21 million light-years away) using an array of radio telescopes.

Another likely candidate for a black hole is the star Cygnus X-1, about six thousand light-years away, which is a massive generator of X-ray radiation. In fact, it is hard to imagine any other physical force besides gravitational collapse that can explain the enormous energy output of stars like Cygnus X-1. Numerous black holes also probably inhabit the center of our own galaxy, a mysterious region with intense radiation and gravitational fields. (If you look at the heavens, the millions of stars comprising our galaxy appear as a faint band of light that cuts across the entire night sky, called the Milky Way. We cannot see the dazzling center because it is hidden from view by dust clouds. However, photographs taken of the centers of neighboring galaxies appear brilliant.)

In the future, scientists will use data from these dead stars to test crucial aspects of general relativity. One physicist who has contributed much to our understanding of the quantum mechanics of black holes is Stephen Hawking, who has struggled against enormous physical handicaps to become a giant in the field of relativity. Hawking, who has lost control of his hands, legs, and mouth, does all his calculations in his head.

Stephen Hawking—Quantum Cosmologist

Some people have declared that Stephen Hawking is the successor to Einstein. He has, in some sense, gone further because he has tried to use quantum mechanics to calculate corrections to the dynamics of black holes. Hawking, by looking at the effects of quantum corrections to the black hole, has predicted phenomena that Einstein never anticipated. He introduced the concept that black holes can "evaporate" and turn into mini-black holes—that is, some light can, in fact, escape the enormous gravitational pull of the black hole because of the Heisenberg Uncertainty Principle, which states that there is a finite though small probability that a light beam can travel against the force of gravity and "leak" past the black hole's enormous grip. This loss of energy from the black hole eventually creates a mini-black hole, which may be as small as a proton.

Hawking's interest in science surfaced when he was young. His father, a medical researcher at the National Institute in London, introduced him to biology at an early age. Hawking recalls:

> I always wanted to know how everything worked. . . . I went through a phase of being very interested in ESP, at about the age of fifteen. A group of us even conducted dice-throwing experiments. Then we heard a lecture by someone who had gone

> through all the famous ESP experiments by Rhine at Duke University. He found that whenever they got results the experimental techniques were faulty, and whenever the experimental techniques were really good they did not get results. So that convinced me that it was all a fraud.[2]

Although talented, Hawking was an indifferent student at Oxford University, lacking the drive and determination that has fired previous great scientists. Then tragedy struck, which changed the course of his life. As a first-year graduate student at Cambridge, he found himself stumbling and slowly losing control over his limbs. He was diagnosed as suffering from Lou Gehrig's disease (amyotrophic lateral sclerosis), an incurable affliction that would inexorably waste away the muscles in his arms and legs.

Today, Hawking has a special mechanical page-turner that allows him to read mathematical equations. Several of his assistants were once specially trained to understand his slow, tortured mumblings, because he had largely lost control of his mouth muscles. It often took him several agonizing seconds to say each word. Nonetheless, he has given learned scientific talks before hundreds of distinguished scientists. A complete invalid, he cruises busily around the Cambridge University campus in his electric wheelchair.

Hawking's desk is littered with mathematical articles sent from colleagues around the world, as well as fan letters from everyone from well-wishers to crackpots trying to sell their latest harebrained ideas. "Being famous is really rather a nuisance,"[3] he once remarked to a journalist.

Hawking states philosophically, "I think I am happier now than I was before I started. Before the illness set on I was very bored with life. I drank a fair amount, I guess; I didn't do any work. When one's expectations are reduced to zero, one really appreciates everything that one does have."[4]

General relativity is one discipline where scientists routinely fill several hundred pages with algebraic equations. Hawking, however, is unique among physicists because he is forced to crank through these calculations in his head. Although he has some help from his students in performing certain calculations, Hawking (like Einstein,

Feynman, and other great scientists) thinks in terms of pictures that express the essential physical concept. The math comes later.

Puzzle of Flatness

In the old framework of Einstein's equations, there were two major problems that had no satisfactory solution. Fortunately, the application of quantum mechanics yields an acceptable solution to them both.

One of the most puzzling features of our universe, as we look around the heavens, is that it seems so flat. This is unusual because, from Einstein's equations, we would expect the universe to have some measurable curvature, either positive or negative.

Second, the universe, no matter where we look, has the same uniform density of galaxies. In fact, if we look at a galaxy, say, a billion light-years in one direction and another billion light-years in another direction, the universe seems pretty much the same. This is curious, because nothing can go faster than the speed of light. How could information about the density of these two galaxies have traveled so far in such a short time? The speed of light, no matter how fast it may appear to us, is too slow to account for the uniform density of the universe across such vast distances.

The answer to these two puzzles was provided by Alan Guth of MIT and improved by Paul Steinhardt at the University of Pennsylvania and Russian physicist A. Linde from Moscow. According to their calculations, when the universe was between 10^{-35} and 10^{-33} seconds old, it underwent an exponential expansion, increasing its radius by a fantastic 10^{50} amount. This "inflation" phase happened just before and was even more rapid than the standard Big Bang phase.

The fact that our universe went through such a massive expansion explains these two puzzles. First of all, our universe seems flat only because the universe is 10^{50} times larger than we once suspected. Think of the analogy with the balloon being blown up. If the balloon is several trillion times larger than before, its surface certainly seems flat.

The inflation scenario also explains the uniformity of the universe. Since, near the beginning of the inflationary period, our entire visible part of the universe was only a tiny speck on the surface of the universe, it was possible for our tiny speck to be mixed uniformly. Inflation simply blew up this uniform speck into our present visible universe. That tiny speck now includes our earth and galaxy, as well as the farthest galaxies that can be seen in our telescopes.

Is Our Universe Unstable?

Although the prospect of a universe 10^{50} times bigger than the known universe is mind-boggling, there is yet another unsettling prospect arising from the GUT and the superstring theories, concerning the catastrophic destruction of our universe.

The ancients often speculated how the earth will end, whether it will be in fire or in ice. The most plausible answer from modern astronomy is that the earth will die in fire, because our sun, once it has used up its hydrogen fuel, will burn unused helium fuel and then expand enormously into a red giant star as large as the orbit of Mars. This means that our earth will be vaporized and we will be roasted in the atmosphere of our sun. All the atoms in our bodies will break up in the sun's atmosphere. (This disaster lies several billion years in the future.)

Moreover, the GUT and superstring theories allow for an even greater disaster than the vaporization of the earth. Physicists predict that matter always tries to search for the lowest-energy state (called the "vacuum state"). For example, as mentioned previously, water always tries to flow downhill. However, this can be changed if we dam up a river. The water that is backed up behind the dam is in a "false-vacuum state," which is not the lowest-energy state. This means that the water would prefer to burst the dam and flow to its true vacuum state below the dam, but it cannot.

Normally, a dam is sufficient to keep water in this false-vacuum state. However, in quantum mechanics, there is always the probability that the water will make a "quantum leap" and penetrate through the dam. According to the Uncertainty Principle, because you don't know where the water is, there is a certain probability that you will find it where you least expect it (that is, on the other side of the dam). Physicists surmise that the water may "tunnel" its way through the barrier.

This leaves us with a rather disturbing thought. Perhaps our entire universe is temporarily sitting in a "false vacuum." What if our universe is not the lowest-energy universe possible? What if another universe exists that has even lower energy, and a sudden quantum transition occurs?

This would be catastrophic. In the new vacuum, the laws of physics and chemistry might be rescrambled beyond recognition. Matter, as we know it, might not even exist, and totally new laws of physics and chemistry could appear. It's often said that the laws of physics are immutable. However, if the universe makes the sudden quantum leap to a lower-vacuum state, then the very laws of physics as we know them also may change beyond recognition.

How would this disaster take place?

A simple visualization of a quantum transition is the boiling of water. Notice that the boiling does not take place instantly, but rather at points, creating bubbles that expand rapidly. Eventually the bubbles coalesce, creating steam. Similarly, if a quantum transition were made to another, lower-energy vacuum, our universe might form "bubbles" that then expand near or at the speed of light (which means that we would never know what hit us). Inside the bubble, there might be alien laws of physics and chemistry. Astronomers could never see this bubble because of its enormous velocity. We might be doing our laundry when suddenly this bubble hits the earth. Suddenly, the very quarks in our bodies might come apart, dissolving us into a chaotic plasma of subatomic particles.

We needn't worry about such a catastrophe, however. Since our universe has been relatively stable for the past 10 to 20 billion years, it is safe to conclude that it has reached the lowest-energy state, although one can never completely rule out the possibility of other universes.

Before the Big Bang

However unsettling the idea is that our entire universe may be unstable, there is one virtue to this idea: It answers the question of what happened before the Big Bang.

As we mentioned before, according to the superstring theory, the universe began in ten dimensions. However, perhaps this ten-dimensional universe was in a false vacuum and therefore unstable. If the ten-dimensional universe did not have the lowest energy, then it would be only a matter of time before it made the "quantum leap" to a lower-energy state.

We now believe that the original expansion of the universe had its origin in a much greater, much more explosive process: the breakdown of the ten-dimensional fabric of space-time. Like a dam bursting, the ten-dimensional fabric of space-time ruptured violently and rapidly re-formed into two separate universes of lower energy: a four-dimensional universe (our own) and a six-dimensional one.

The violence of this explosion could easily generate enough energy to drive the inflation process forward. The standard Big Bang expansion would emerge only later, as the inflation process slowed down and made the transition to a traditional expanding universe.

The four-dimensional universe expanded at the expense of the six-dimensional universe, which collapsed down to the Planck length. That is perhaps the reason why our universe appears to be four-dimensional—the other six dimensions, although they are all around us, are too small to be observed.

Although we are still far from being able to verify this picture experimentally, the rapidly developing field of cosmology has given us tantalizing clues about the nature of matter. Some physicists feel that the answer to many of our questions about the universe may lie in the substance called "dark matter," perhaps the most mysterious form of matter in the universe.

10

The Mystery of Dark Matter

WITH THE CANCELLATION of the SSC, some commentators have publicly speculated that physics will "come to an end." Promising ideas such as the superstring theory, no matter how compelling and elegant, will never be tested and, hence, can never be verified. Physicists, however, are optimists. If evidence for the superstring theory cannot be found on the earth, then one solution is to leave the earth and go into outer space. Over the coming years, physicists will rely increasingly on cosmology to probe the inner secrets of matter and energy. Their laboratory will be the cosmos and the Big Bang itself.

Already, cosmology has given us several mysteries that may very well provide clues to the ultimate nature of matter. The first is dark matter, which makes up 90 percent of the universe. And the second is cosmic strings, which we will discuss in the next chapter.

What Is the World Made Of?

One of the greatest achievements of twentieth-century science was the determination of the chemical elements of the universe. With only a little over one hundred elements, scientists could explain the trillions upon trillions of possible forms of matter, from DNA to animals to exploding stars. The familiar elements that made up the earth—such as carbon, oxygen, and iron—were the same as the elements making up the distant galaxies. Analyzing the light taken from blazing stars billions of light-years from our galaxy, scientists found precisely the same familiar elements found in our own backyards, no more, no less.

Indeed, no new mysterious elements were found anywhere in the universe. The universe was made of atoms and their subatomic constituents. That was the final word in physics.

But by the late twentieth century, an avalanche of new data has confirmed that over 90 percent of the universe is made of an invisible form of unknown matter, or dark matter. The stars we see in the heavens, in fact, are now known to make up only a tiny fraction of the real mass of the universe.

Dark matter is a strange substance, unlike anything ever encountered before. It has weight but cannot be seen. In theory, if someone held a clump of dark matter in their hand, it would appear totally invisible. The existence of dark matter is not an academic question, because the ultimate fate of the universe, whether it will die in a fiery Big Crunch or fade away in a Cosmic Whimper or Big Chill, depends on its precise nature.

High-mass subatomic vibrations predicted by the superstring theory are a leading candidate for dark matter. Thus, dark matter may give us an experimental clue to probe the nature of the superstring. Even without the SSC, science may be able to explore the new physics beyond the Standard Model.

How Much Does a Galaxy Weigh?

The scientist who first suspected that there was something wrong about our conception of the universe was Fritz Zwicky, a Swiss-American astronomer at the California Institute of Technology. In the 1930s, he was studying the Coma cluster of galaxies, about 300 million light-years away, and was puzzled by the fact that they were revolving about each other so fast that they should be unstable. To confirm his suspicions, he had to calculate the mass of a galaxy. Since galaxies can contain hundreds of billion stars, calculating their weight is a tricky question.

There are two simple ways of making this determination. The fact that these two methods gave startlingly different results has created the present crisis in cosmology.

First, we can count the stars. This may seem like an impossible task, but it's really quite simple. We know the rough average density of the galaxy, and then we multiply by the total volume of the galaxy. (That's how, for example, we calculate the number of hairs on the human head, and how we determine that blondes have fewer hairs than brunettes.)

Furthermore, we know the average weight of the stars. Of course, no one actually puts a star on a scale. Astronomers instead look for binary star systems, where two stars rotate around each other. Once we know the time it takes for a complete rotation, Newton's laws are then sufficient to determine the mass of each star. By multiplying the number of stars in a galaxy by the average weight of each star, we get a rough number for the weight of the galaxy.

The second method is to apply Newton's laws directly on the galaxy. Distant stars on a spiral arm of the galaxy, for example, orbit around the galactic center at different rates. Furthermore, galaxies and clusters of stars rotate around each other. Once we know the time it takes for these various revolutions, we can then determine the total mass of the galaxy using Newton's laws of motion.

Zwicky calculated the mass necessary to bind this cluster of galaxies by analyzing the rate at which they orbited around each other. He found that this mass was twenty times greater than the actual mass of the luminous stars. In a Swiss journal, Zwicky reported that there was a fundamental discrepancy between these two results. He postulated that there had to be some form of mysterious ''dunkle Materie,'' or dark matter, whose gravitational pull held this galactic cluster together. Without this dark matter, the Coma galaxies should fly apart.

Zwicky was led to postulate the existence of dark matter because of his unshakable belief that Newton's laws were correct out to galactic distances. (This is not the first time that scientists predicted the presence of unseen objects based on faith in Newton's laws. The planets Neptune and Pluto, in fact, were discovered because the orbit of closer planets, such as Saturn, wobbled and deviated from Newton's predictions. Rather than give up Newton's laws, scientists simply predicted the existence of new outer planets.)

However, Zwicky's results were met with indifference, even hostility, by the astronomical community. After all, the very existence of galaxies beyond our own Milky Way galaxy had been determined only nine years before by Edwin Hubble, so most astronomers were convinced that his results were premature, that eventually they would fade away as better, more precise observations were made.

So Zwicky's results were largely ignored. Over the years, astronomers accidentally rediscovered them but dismissed them as an aberration. In the 1970s, for example, astronomers using radio telescopes analyzed the hydrogen gas surrounding a galaxy and found that it rotated much faster than it should have, but discounted the result.

In 1973, Jeremiah Ostriker and James Peebles at Princeton University resurrected this theory by making rigorous theoretical calculations about the stability of a galaxy. Up to that time, most astronomers thought that a galaxy was very much like our solar system, with the inner planets traveling much faster than the outer planets. Mercury, for example, was named after the Greek god for speed since it raced across the heavens (traveling at 107,000 miles per hour). Pluto, on the other hand, lumbers across the solar system at 10,500 miles per hour. If Pluto traveled around the sun as fast as Mercury, then it would quickly fly into outer space, never to return. The gravitational pull of the sun would not be enough to hold on to Pluto.

However, Ostriker and Peebles showed that the standard picture of a galaxy, based on our solar system, was unstable; by rights, the galaxy should fly apart. The gravitational pull of the stars was not enough to hold the galaxy together. They then showed that a galaxy can become stable if it is surrounded by a massive invisible halo that holds the galaxy together and if 90 percent of its mass was actually in the halo in the form of dark matter. Their paper was also met with indifference.

But after decades of skepticism and derision, what finally turned the tide on dark matter was the careful, persistent results of astronomer Vera Rubin and her colleagues at the Carnegie Institution in Washington, D.C. The results of these scientists, who analyzed hundreds of galaxies, verified conclusively that the velocity of the outer stars in a galaxy did not vary much from that of the inner ones, contrary to the planets in our solar system. This meant that the outer stars should fly into space, causing the galaxy to disintegrate into billions of individual stars, unless held together by the gravitational pull of invisible dark matter.

Like the history of dark matter itself, it took several decades for Vera Rubin's lifetime of results to be recognized by the skeptical (and overwhelmingly male) astronomical community.

One Woman's Challenge

It has never been easy for a female scientist to be accepted by her male peers. In fact, at every step of the way, Dr. Rubin's career came perilously close to being derailed by male hostility. She first became interested in the stars in the 1930s as a ten-year-old child, gazing at the night sky over Washington, D.C., for hours at a time, even making detailed maps of meteor trails across the heavens.

Her father, an electrical engineer, encouraged her to pursue her interest in the stars, even helping her build her first telescope at the age of fourteen and taking her to amateur astronomy meetings in Washington. However, the warm encouragement she felt inside her family contrasted sharply with the icy reception she received from the outside world.

When she applied to Swarthmore College, the admissions officer tried to steer her away from astronomy, to a more "ladylike" career of painting astronomical subjects. That became a standard joke around her family. She recalled, "Whenever anything went wrong for me at work, someone would say, 'Have you ever thought of a career in which you paint? . . .' "[1]

When accepted at Vassar, she proudly told her high school physics teacher in the hallway, who replied bluntly, "You'll do all right as long as you stay away from science." (Years later, she recalled, "It takes an enormous amount of self-esteem to listen to things like that and not be demolished."[2])

After graduating from Vassar, she applied to graduate school at Princeton, which had a world-renowned reputation in astronomy. However, she never even received the school's catalog, since Princeton did not accept female graduate students in astronomy until 1971.

She was accepted at Harvard, but declined the offer because she had just gotten married to Robert Rubin, a physical chemist, and followed him to Cornell University, where the astronomy department consisted of just two faculty members. (After she declined, she got a formal letter back from Harvard, with the handwritten words scrawled on the bottom, "Damn you women. Every time I get a good one ready, she goes off and gets married."[3])

Going to Cornell, however, was a blessing in disguise, since Rubin took graduate courses in physics from two Nobel laureates in physics, Hans Bethe, who decoded the complex fusion reactions which energize the stars, and Richard Feynman, who renormalized quantum electrodynamics. Her master's thesis met head-on the hostility of a male-dominated world. Her paper, which showed that the faraway galaxies deviated from the uniform expansion of a simplified version of the Big Bang model, was rejected for publication because it was too far-fetched for its time. (Decades later, her paper would be considered prophetic.)

But after receiving her master's degree from Cornell, Rubin found herself an unhappy housewife. "I actually cried every time the *Astrophysical Journal* came into the house . . . nothing in my education had taught me that one year after Cornell my husband would be out doing his science and I would be home changing diapers."[4]

Nonetheless, Rubin struggled to pursue her childhood dream, especially after her husband took a job in Washington. Taking nighttime classes, she received her Ph.D. from Georgetown University. In 1954, she published her Ph.D. thesis, a landmark study that showed that the distribution of the galaxies in the heavens was not smooth and uniform, as previously thought, but actually clumpy.

Unfortunately, she was years ahead of her time. Over the years, she gained a reputation of being something of an eccentric, going against the prevailing prejudice of astronomical thought. It would take years for her ideas to gain the recognition they deserved.

Distressed by the controversy her work was generating, Rubin decided to take a respite and study one of the most mundane and unglamorous areas of astronomy, the rotation of galaxies. Innocently enough, Rubin began studying the Andromeda galaxy, our nearest neighbor in space. She and her colleagues expected to find that the gas swirling in the outer fringes of the Andromeda galaxy was traveling much slower than the gas near the center. Like our own solar system, the speed of the gas should slow down as one went farther from the galactic nuclei.

Much to their surprise, they found that the velocity of the gas was a constant, whether it was near the center or near the rim of the galaxy. At first, they thought this peculiar result was unique to the Andromeda galaxy. Then they systematically began to analyze hundreds of galaxies (two hundred galaxies since 1978) and found the same curious result. Zwicky had been right all along.

The sheer weight of their observational results could not be denied. Galaxy after galaxy showed the same, flat curve. Because astronomy had become technically much more sophisticated since the time of Zwicky, it was possible for other laboratories to verify Rubin's numbers rapidly. The constancy of velocity of a rotating galaxy was now a universal fact of galactic physics. Dark matter was here to stay.

For her pioneering efforts, Vera Rubin was elected to the prestigious National Academy of Science in 1981. (Since it was founded in 1863, only 75 women among the 3,508 scientists have been elected to the academy.)

Today, Rubin is still pained by how little progress female scientists have made. Her own daughter has a Ph.D. in cosmic ray physics. When she went to Japan for an international conference, she was the only woman there. "I really couldn't tell that story for a long time without weeping," Rubin recalled, "because certainly in one generation, between her generation and mine, not an awful lot has changed."[5]

Not surprisingly, Rubin is interested in stimulating the interest of young girls to pursue scientific studies. She has even written a children's book, entitled *My Grandmother Is an Astronomer.*

Bending Starlight

Since Rubin's original paper, even more sophisticated analyses of the universe have shown the existence of the dark matter halo, which may be as much as six times the size of the galaxy itself. In 1986, Bodhan Paczynski of Princeton University realized that if the starlight from a distant star traveled by a nearby clump of dark matter, the dark matter might bend the starlight and act as a magnifying lens, making the star appear much brighter. In this way, by looking for dim stars that suddenly got brighter, the presence of dark matter could be detected. In 1994, two groups independently reported photographing such a stellar brightening. Since then, other teams of astronomers have joined in, hoping to find more examples of stellar brightening.

In addition, the bending of starlight by a distant galaxy can be used as another way in which to calculate the galaxy's weight. Anthony Tyson and his colleagues at the AT&T Bell Laboratories have analyzed light rays from dim blue galaxies at the rim of the visible universe. This cluster of galaxies acts like a gravitational lens, bending the light from other galaxies. Photos of distant galaxies have confirmed that the bending is much more than expected, meaning that their weight comes from much more than the sum of their individual stars. Ninety percent of the mass of these galaxies turns out to be dark, as predicted.

Hot and Cold Dark Matter

While the existence of dark matter is no longer in dispute, its composition is a matter of lively controversy. Several schools of thought have emerged, none of them very satisfactory.

First, there is the "hot dark matter" school, which holds that dark matter is made of familiar lightweight subparticles such as neutrinos, which are notoriously difficult to detect. Since the total flux of neutrinos filling up the universe is not well known, the universe may be bathed in a flood of neutrinos, making up the dark matter of the universe.

If the electron-neutrino, for example, is found to have a tiny mass, then there is a chance that it may have enough mass to make up the missing mass problem. (In February 1995, physicists at the Los Alamos National Laboratory in New Mexico announced that they had found evidence that the electron-neutrino has a tiny mass: one-millionth the weight of an electron. However, this result must still be verified by other laboratories before it is finally accepted by other physicists.)

Then there is the "cold dark matter" school, which suspects that dark matter is made of heavier, slow-moving, and much more exotic subparticles. For the past decade, physicists have been looking for exotic candidates that might make up cold dark matter. These particles have been given strange, whimsical names, such as "axions," named after a household detergent. Collectively, they are called WIMPs, for "weakly interacting massive particles." (The skeptics have retaliated by pointing out that a significant part of dark matter may consist of familiar but dim forms of ordinary matter, such as red dwarf stars, neutron stars, black holes, and Jupiter-sized planets. Not to be outdone, they have called these objects MACHOs, for "massive astrophysical compact halo objects." However, even the proponents of MACHOs admit that, at best, they can explain only 20 percent of the dark matter problem. In late 1994, however, a version of the MACHO theory was dealt a blow when the Hubble Space Telescope, scanning the Milky Way galaxy for red dwarf stars, found far fewer of these dim stars than expected.)

But perhaps the most promising candidate for WIMPs are the super particles, or "sparticles" for short. Supersymmetry, we remember, was first seen as a symmetry of particle physics in the superstring theory. Indeed, the superstring is probably the only fully consistent theory of superparticles.

According to supersymmetry, every particle must have a superpartner, with differing spin. The leptons (electrons and neutrinos) for example, have spin ½. Their superpartners are called "sleptons" and have spin 0. Likewise, the superpartners of the quarks are called "squarks" and also have spin 0.

Furthermore, the superpartner of the spin 1 photon (which describes light) is called the "photino." And the superpartner of the gluons (which holds the quarks together) is called "gluino."

The main criticism of sparticles is that we have never seen them in the laboratory. At present, there is no evidence that these superparticles exist. However, it is widely believed that this lack of evidence is only because our atom smashers are too feeble to create superparticles. In other words, their mass is simply too large for our atom smashers to produce them.

Lack of concrete evidence has not, however, prevented physicists from trying to use particle physics to explain the mysteries of dark matter and cosmology. For example, one of the leading candidates for the WIMP is the photino.

The cancellation of the SSC, therefore, does not necessarily doom our attempts to verify the correctness of the superstring. Within the next decade, it is hoped that the increased accuracy of our astronomical observations, with the deployment of a new generation of telescopes and satellites, may narrow down the candidates for dark matter. If dark matter turns out to be composed, at least in part, of sparticles, belief in the superstring theory would receive an enormous boost.

How Will the Universe Die?

Last, dark matter may prove decisive in understanding the ultimate fate of the universe. One persistent controversy has been the fate of an expanding universe. Some believe that there is enough matter and gravity to reverse its expansion. Others believe that the universe is too low in density, so that the galaxies will continue their expansion, until temperatures around the universe approach absolute zero.

At present, attempts to calculate the average density of the universe show the latter to be true: The universe will die in a Cosmic Whimper or a Big Chill, expanding forever. However, this theory is open to experimental challenges. Specifically, there might be enough missing matter to boost the average density of the universe.

To determine the fate of the universe, cosmologists use the parameter called "omega," which measures the matter density of the universe. If omega is greater than one, then there is enough matter in the universe to reverse the cosmic expansion, and the universe will begin to collapse until it reaches the Big Crunch.

However, if omega is less than one, then the gravity of the universe is too weak to change the cosmic expansion, and the universe will expand forever, until it reaches the near-absolute-zero temperatures of the Cosmic Whimper. If omega is equal to one, then the universe is balanced between these two scenarios, and the universe will appear to be perfectly flat, without any curvature. (For omega to equal one, the density of the universe must be approximately three hydrogen atoms per cubic meter.) Current astronomical data favors a value of .1 for omega, which is too small to reverse the cosmic expansion.

The leading modification of the Big Bang theory is the inflationary universe, which predicts a value of omega of precisely 1. However, the visible stars in the heavens only give us 1 percent of the critical density. This is sometimes called the ''missing mass'' problem. (It is different from the dark matter problem, which was based on purely galactic considerations.) Dust, brown dwarfs, and nonluminous stars may boost this number a bit, but not by much. For example, the results from nucleosynthesis show that the maximum value of the density of this form of nonluminous matter cannot exceed 15 percent of the critical density.

Even if we add in the dark matter halos that surround the galaxies, this only brings us up to 10 percent of the critical value. So the dark matter in halos cannot solve the missing mass problem by itself.

In addition to the still-unsolved dark matter problem, there is an equally perplexing cosmological puzzle involving the unexpected clumping of galaxies into gigantic clusters. The solution of this problem may involve another kind of string theory, called ''cosmic strings.''

11

Cosmic Strings

SUPERSTRINGS, of course, are thought to be incredibly tiny vibrating objects, much too small to be seen or detected with our puny instruments. However, some physicists have conjectured that, soon after the Big Bang, there may have been gigantic cosmic strings floating in space, even larger than the galaxies themselves. (These cosmic strings were inspired by superstrings and use many of the same equations, although they are physically different objects.)

According to this theory, the vibrations left in the wake of these ancient wriggling cosmic strings are the galactic clusters we see in the heavens, including our own. Normally, speculations about fantastic cosmic strings stretching across billions of light-years of space may sound like sheer fantasy. However, cosmic strings are quite practical; they may one day explain one of the thorniest puzzles in cosmology, the ''clumpiness'' of the universe. If verified, then our own solar system may owe its existence to these cosmic strings.

(Cosmic strings featured prominently in the movie *Star Trek: Generations*. The plot revolves around a gigantic cosmic string that is seen roaming through the galaxy; its colossal energy field wrecks Federation star ships, creates interstellar havoc, and even captures Captains James Kirk and Jean-Luc Picard in a time warp. And anyone landing on the surface of the cosmic string is transported instantly into a heavenlike dream world, where all wishes come true.)

Why Is the Universe So Clumpy?

In the 1970s, astrophysicists were confident that the Big Bang theory could account for many of the qualitative and quantitative features of the universe. Then some disturbing results were discovered while analyzing the large-scale distribution of galaxies. These new results did not overthrow the Big Bang theory, but they forced cosmologists to revise their understanding.

There were two conflicting results. Data from the cosmic microwave radiation that pervades the universe (the "echo from the Big Bang") indicated that the Big Bang was uniform in all directions. No matter where scientists pointed their instruments, they found that the cosmic microwave radiation was smooth and uniform, filling up the universe with an even background radiation. In fact, sensors launched on balloons, small rockets, and ground-based instruments were unable to find any ripples in this background radiation.

Since the cosmic microwave background radiation dates back to when the universe was only 300,000 years old, these results meant that the echo of the Big Bang was very smooth at that time. (Three hundred thousand years after the Big Bang, the universe had cooled down to the point where stable atoms could form without being ripped apart by the heat. Then leftover radiation from the Big Bang could travel through space without being absorbed by atoms. This leftover radiation forms the microwave background that we see today.)

But when astronomers plotted the location of thousands of galaxies, they found that they clumped together in unusual formations. In fact, there were gigantic voids between galaxies, stretching across millions of light-years. These voids and clumps, spanning distances on the scale of several hundred million light-years, may comprise as much as 5 percent of the visible universe. This clumping began about 100 million years after the Big Bang.

This clumpiness in the distribution of galaxies was first observed by Harlow Shapely of Harvard in 1933 and Fritz Zwicky in 1938. But their data was crude and insufficient to determine if this was a universal phenomenon. Only decades later, with the use of advanced computers, digital analysis, and automatic techniques, were astronomers able to analyze systematically thousands of galaxies at a time. These automatic techniques have produced brilliant galactic charts containing over ten thousand galaxies apiece, showing clearly that the galaxies are not spread evenly throughout the heavens, but clump together, leaving huge voids between them.

In 1987, Adam Dressler of the Carnegie Institution and six collaborators found a large group of galaxies moving together within about 200 million light-years of the earth, as if they were being attracted by an unseen, gigantic mass, which these scientists dubbed the Great Attractor. They found that, in addition to the expansion of the universe, the Milky Way galaxy and other nearby galaxies seem to be rushing toward the Great Attractor, in the direction of the constellation Leo.

In 1989, Margaret Geller and John Huchra of Harvard University announced the discovery of a giant ''wall'' of galaxies stretching about 500 million light-years across. They dubbed it the Great Wall. Then astronomers turned their attention to the Southern Hemisphere, looking for evidence of clumping. Nearly 3,600 galaxies were charted, revealing the existence of a Southern Wall. Later, Tod Lauer of the National Optical Astronomy Observatories in Tucson and Marc Postman of the Space Telescope Science Institute in Baltimore attempted to check these strange results by analyzing a volume of space thirty times larger than that analyzed by Dressler's group. They analyzed 119 galactic clusters up to 500 million light-years away and were surprised to find that the Great Attractor was a small part of an even larger motion. This larger group of galaxies was traveling at roughly 700 kilometers per second toward the constellation Virgo.

The clumpiness of the universe created a puzzle for the Big Bang. If the universe was perfectly smooth about 300,000 years after the Big Bang, then there wasn't enough time for the galaxies to clump together about 100 million years later. Most cosmologists agreed that this was too short a time for an astonishingly smooth explosion to create the large clumping of galaxies.

"The Face of God"

This riddle remained one of the great puzzles of cosmology, until the COBE satellite was launched into space in 1989 to give the first comprehensive picture of the cosmic microwave background. Finally, cosmologists had a wealth of information about the precise characteristics of the background radiation.

The goal of the COBE satellite was to find tiny temperature variations within the otherwise smooth microwave background radiation. These tiny kinks and hot spots would serve as "seeds" that eventually would grow into the clumps of galaxies that we see today. If the COBE satellite did not find any tiny anomalies, then our understanding of the evolution of the universe had to be revised significantly.

It took many months for the physicists at Berkeley to sift through the COBE data. Most important, they had to subtract all extraneous effects, such as static coming from the Milky Way galaxy, and even the motion of the earth and sun with respect to the cosmic microwave background.

The results hit the front pages of every major newspaper in the world in April 1992. Careful analysis showed that the cosmic microwave background radiation was not so uniform and that tiny irregularities showed up that were too small to be detected by previous experiments. Photos of these irregularities were given to reporters, with the tongue-in-cheek statement that they were gazing at the face of God.

According to the revised theory, these small anomalies in the microwave radiation existed 300,000 years after the Big Bang and grew in size over the next billion years, until they caused galaxies to clump randomly in space. Calculations showed that these small perturbations in the otherwise smooth radiation were sufficient to cause the clumpiness. As the universe expanded, these tiny kinks in the radiation also grew in size, until they gradually became the clumps of galaxies that we see today.

Dark Matter and the Clumpiness of the Universe

Perhaps the simplest explanation of the clumpiness of the universe comes from the dark matter theory. We recall that the cosmic microwave radiation represents the remnants of radiation dating 300,000 years after the Big Bang. Before that, ordinary matter was simply too hot to form any clumping. Any atoms that tried to clump together would be torn apart by the intense heat. Astronomer Donald Goldsmith compares this to trying to make a soufflé in the midst of a tornado. As soon as you bake one in the oven, it is torn apart by the fierce winds. However, there is no problem if you wait until the winds die down.

Dark matter is an exception to this picture. Clumping may have taken place much earlier than 300,000 years after the Big Bang if there were significant amounts of dark matter. Since dark matter does not interact with ordinary electromagnetic radiation, it would not be affected by the intense radiation fields that existed less than 300,000 years after the Big Bang. But dark matter does have gravity, so clumps of dark matter might have begun to form almost immediately after the Big Bang. Then, after 300,000 years, ordinary matter would be attracted to these large clumps of dark matter, forming the galaxies and galactic clusters that we see today with our telescopes.

However, there is yet another way in which to explain the clumping of the universe, and this is through cosmic strings.

Topological Defects

Our understanding of cosmic strings comes from our understanding of ordinary phase transitions, which are found in everything from the formation of crystals to the magnetization of iron. For example, solid state physicists realize that phase transitions (such as melting, freezing, and boiling) are not smooth, uniform transitions but abrupt events that begin with the formation of microscopic "defects" in the atomic structure of matter that grow rapidly.

When a phase transition is about to occur, these tiny defects arise like fault lines in the atomic lattice array and have definite physical shapes, such as lines and walls. Microscopic photographs taken of ice as it begins to freeze show that freezing occurs when tiny linelike and wall-like defects arise, which then act as "seeds" around which tiny ice crystals form.

Likewise, in microscopic pictures of iron as it is placed in a magnetic field, one can see tiny "walls" begin to form among the atoms. Within each "domain" separated by these walls, the iron atoms are pointed in a certain direction. As the magnetic field increases, these walls merge, until all the atoms are pointing in the same magnetic direction.

Particle physicists believe that similar defects occurred in the early universe, when the Big Bang began to cool down. As the early subatomic particles began to cool, they might have condensed into these defects, which include strings, walls, and more complicated structures called "textures."

These ancient cosmic strings have an analogy in ordinary magnetism. Magnetic fields normally cannot penetrate superconducting material (which has zero electrical resistance and is cooled to near-absolute-zero temperatures). However, magnetic fields can penetrate certain types of superconductors and form strings of condensed magnetic fields. Thus, magnetic fields, instead of being pervasive, are concentrated into thin strings that penetrate this superconductor.

Similarly, cosmic strings may be likened to condensed subatomic fields from the early universe. They have no ends; they are either closed or are infinitely long. According to this scenario, these one-dimensional fault lines were formed soon after the Big Bang began to cool and condensed into a web of tangled strings threading the entire universe.

These strings have enormous tension in them, so they vibrate and wiggle violently, often intersecting other strings themselves. Originally, it was thought that cosmic strings were several hundred thousand light-years across and therefore formed the seeds out of which the galaxies grew. However, computer programs simulating the growth of cosmic strings seem to rule this out.

In the 1980s, it was proposed that cosmic strings, by their violent motion, create wakes of "gravity waves," just as motor boats create crests of waves as they speed across a lake. This wall of gravity waves later would condense into sheetlike formations of matter, similar to the walls of galaxies found today. If the primeval universe had a magnetic field, then it could be shown that these cosmic strings could have generated enormous electrical fields, becoming superconductors themselves. As these superconducting cosmic strings thrashed about in the early universe, they might have pushed about matter rather than attracting it. Either way, we would find irregularities in the distribution of matter.

These puzzles may be resolved with the next generation of experiments and observations. Previous maps of the universe have recorded the precise positions and velocities of tens of thousands of galaxies at a time. Donald G. York, of the University of Chicago, is heading a group effort of several universities to collect perhaps the largest registry of galaxies, up to one million galaxies, starting in 1995. Advances in automated and digitalized optical instruments are making this previously unimaginable feat possible. Such a galactic atlas would go a long way toward determining if these anomalies persist.

Perhaps the most important set of experiments would involve refinements on the COBE satellite data. One of the limitations of the COBE satellite was that it could analyze temperature variations only down to 7 degrees of arc. (This arc covers an area of sky blocked out by a large grapefruit held out at arm's length.) Fluctuations over such large regions of space correspond to ten times the size of the largest superclusters of galaxies seen today. This means that the COBE satellite was incapable of detecting tiny temperature variations spread over areas that eventually became the recently observed galactic clusters.

Unfortunately, balloon experiments do not last long enough to make reliable measurements, and ground-based sensors are limited by fluctuations in the atmosphere. Ultimately, future cosmological experiments lie in another COBE-like satellite capable of detecting temperature variations within .5 degrees of arc.

Dark matter, cosmic strings, and galactic clumping are some of the issues that will continue to intrigue and baffle cosmologists for years to come. Because these concepts can be measured and tested, we hope within this decade to resolve many of these experimental questions. With the cancellation of the SSC, we will rely increasingly on this ever-expanding body of cosmological information to probe the limits of the Standard Model and beyond. It is hoped we may even gain a glimpse into perhaps the most fascinating aspect of superstring theory, higher-dimensional space-time. To better understand how the universe was born, we will now investigate ten dimensional space and time.

12

Journey to Another Dimension

BACK IN 1919, when Einstein was still engrossed in calculating the consequences of his new theory of general relativity, he received a letter from an unknown mathematician, Theodor Franz Kaluza, from the University of Konigsberg (in what is now the city of Kaliningrad in the Russian federation).

In this letter, Kaluza proposed a novel way of writing a unified field theory that combined Einstein's new theory of gravity with the older theory of light written by Maxwell. Instead of describing a theory with three space dimensions and one time dimension, Kaluza proposed a five-dimensional theory of gravity. With this fifth spatial dimension, Kaluza had enough room to fit the electromagnetic force into Einstein's theory of gravity. In one stroke, it seemed that Kaluza had provided a fundamental clue to the problem that Einstein was working on. Kaluza did not have the slightest experimental proof that the world should be five-dimensional, but his theory was so elegant that it seemed it might have some truth.

The idea of five dimensions was so outlandish to Einstein that he held on to the paper, delaying its publication for two years. However, instinct told Einstein that the mathematics of this theory were so beautiful that it might just be correct. In 1921, Einstein finally gave his approval to the Prussian Academy to publish Kaluza's paper.

In April 1919, Einstein wrote to Kaluza, "The idea of achieving [a unified field theory] by means of a five-dimensional cylinder world never dawned on me. . . . At first glance I like your idea enormously."[1] A few weeks later, Einstein wrote again, "The formal unity of your theory is startling."[2]

However, most physicists viewed the Kaluza theory with skepticism. They had a hard time understanding Einstein's four dimensions, let alone Kaluza's five. Furthermore, Kaluza's theory raised more questions than it answered. If the unification of light with gravity requires five dimensions, but only four can be measured in our laboratories, then what happened to the fifth dimension?

To some physicists, this new theory seemed to be a parlor trick, devoid of physical content. However, physicists like Einstein realized that this discovery was so simple and elegant that it might be of first rank. The trouble with all this was: What did it mean?

Indeed, it was preposterous to suggest that the world was five-dimensional. For example, if a bottle of gas is opened and placed in a sealed room, sooner or later the gas molecules, by random collisions, will seek out and diffuse into all possible spatial dimensions. However, it is obvious that these gas molecules will fill up only three dimensions.

So where did the fifth dimension go? Einstein felt that Kaluza's trick was simply too good to throw away just on the grounds that it violated intuition about our known universe. Once again, beauty alone, without experimental verification, was sufficient grounds for Einstein to consider the theory seriously. Finally, in 1926, Swedish mathematician Oskar Klein discovered a possible solution to the problem.

Kaluza had suggested earlier that the fifth dimension was different from the other four dimensions because it was "curled up," like a circle. To explain why the universe obviously appears to be four-dimensional, Klein suggested that the size of this circle was so small that it could not be observed directly.

In other words, the gas molecules released in a room will indeed seek out all possible spatial dimensions, but the gas molecules are simply too big to fit into the circular fifth dimension. As a consequence, the gas molecules fill up only four dimensions.

Klein even calculated a possible size of the fifth dimension: the

Planck length, which is (10^{-33} cm), or about a hundred billion billion times smaller than the nucleus of an atom.

Klein's brilliant solution to where the fifth dimension went also raised more questions than it solved. For example, why did the fifth dimension ball up into a small circle, leaving the other dimensions extending out to infinity?

Einstein would struggle for the next thirty years to make sense out of the Kaluza-Klein theory, as it was called, as a candidate for the unified field theory, but he could not solve this puzzling question.

In the latter half of his life, Einstein would work in two avenues: The first was his own geometric version of electromagnetism, which described the light force as a simple distortion of the fabric of space-time. This avenue led to more complicated mathematics, and was ultimately a dead end. The second was the Kaluza-Klein theory, which was beautiful pictorially but useless as a model of our universe. The theory had great promise, if only someone could figure out why the fifth dimension curled up. Einstein worked on the Kaluza-Klein theory periodically but made no progress.

The Solution: Quantum Strings

For the next fifty years, most physicists left the ideas of Kaluza and Klein on the shelf, considering them a curious footnote to the bizarre nature of pure mathematics. The theory was almost forgotten until the 1970s, when Scherk wondered whether the Kaluza-Klein trick of curling up the unwanted dimensions could solve his problem. He and his colleague, E. Cremmer, proposed this as the solution to the problem of going from twenty-six or ten dimensions down to four dimensions.

The superstring physicists, however, had one great advantage over Kaluza and Klein: They could use the full power of quantum mechanics developed over the decades to solve the problem of why the higher dimensions curled up.

Previously, we learned that quantum mechanics makes possible the phenomenon of symmetry breaking. Nature always prefers the state of lowest energy. Although our original universe may have been symmetrical, it also may have been in a higher-energy state and therefore would make a ''quantum leap'' to a lower-energy state.

Similarly, it is believed that the original ten-dimensional string was unstable; it was not the state that possessed the lowest energy.

Today theoretical physicists are making intense efforts to prove that the lowest-energy state predicted by the superstring model is a universe in which six dimensions have curled up, leaving our four-dimensional universe intact. The current belief is that the original ten-dimensional universe was actually a false vacuum—that is, it was not the state of lowest energy.[3]

Although no one has succeeded in proving that the ten-dimensional universe was unstable and made the quantum leap to four dimensions, physicists are optimistic that the theory is rich enough to allow for this possibility. For young scientists trying to solve the important contemporary problems of physics, therefore, one of the most outstanding problems of the superstring theory is to show conclusively that the ten-dimensional universe made the quantum leap to our known four-dimensional universe.

Mr. Square

In science fiction novels, a trip into higher dimensions resembles a journey into a strange but earthlike world. In these novels, people are similar to ourselves, but with some twist. This common misconception is due to the fact that the imaginations of science fiction writers are too limited to grasp the true features of higher-dimensional universes given to us by rigorous mathematics. Science is truly stranger than science fiction.

The simplest way to understand higher-dimensional universes is to study lower-dimensional universes. The first writer to undertake this task in the form of a popular novel was Edwin A. Abbott, a Shakespearean scholar who in 1884 wrote *Flatland,* a Victorian satire about the curious habits of people who live in two spatial dimensions.

Imagine the people of Flatland living, say, on the surface of a table. This tale is narrated by the pompous Mr. A. Square, who tells us of a world populated by people who are geometric objects. In this stratified world, the women are Straight Lines, workers and soldiers are Triangles, professional men and gentlemen (like himself) are Squares, and the nobility are Pentagons, Hexagons, and Polygons.

The more sides on a person, the higher his social rank. Some noblemen have so many sides that they eventually become Circles, which is the highest rank of all.

Mr. Square, a man of considerable social rank, is content to live in the pampered tranquility of this ordered society until one day strange beings from Spaceland (a three-dimensional world) appear before him and introduce him to the wonders of another dimension.

When the people of Spaceland look at Flatlanders, they can see inside their bodies and view their internal organs. This means that the people of Spaceland, in principle, can perform surgery on the people of Flatland without cutting their skin.

What happens when higher-dimensional beings enter a lower-dimensional universe? When the mysterious Lord Sphere of Spaceland enters Flatland, Mr. Square can only see circles of ever-increasing size penetrate his universe. Mr. Square cannot visualize Lord Sphere in his entirety, only cross-sections of his shape.

Lord Sphere even invites Mr. Square to visit Spaceland, which involves a harrowing journey where Mr. Square is peeled off his Flatland world and deposited in the forbidden third dimension. However, as Mr. Square moves in the third dimension, his eyes can see only two-dimensional cross-sections of the three-dimensional Spaceland. When Mr. Square meets a Cube, he sees it as a wondrous object that appears as a square within a square that constantly changes shape as he looks at it.

Mr. Square is so shaken by his encounter with the Spacelanders that he decides to tell his fellow Flatlanders of his remarkable journey. His tale, which might upset the ordered society of Flatland, is perceived as seditious by the authorities, and he is arrested and brought before the Council. At his trial, he tries, in vain, to explain the third dimension. To the Polygons and the Circles, he tries to explain the three-dimensional Sphere, the Cube, and the world of Spaceland.

Mr. Square is sentenced to perpetual imprisonment in jail (which consists of a line drawn around him) and lives out his life as a martyr. (Ironically, all Mr. Square has to do is "jump" out of the prison into the third dimension, but this is beyond his comprehension.)

Mr. Abbott, a theologian and headmaster of the City of London School, wrote *Flatland* as a political satire on the Victorian hypocrisy he saw around him. However, one hundred years after he wrote *Flatland,* the superstring theory requires physicists to think seriously of what a higher-dimensional universe might look like.

First of all, a ten-dimensional being looking down on our universe could see all of our internal organs and could even perform surgery on us without cutting our skin. This idea of reaching into a solid object without breaking the outer surface seems absurd to us only because our minds are limited when considering higher dimensions, just like the minds of the Polygons on the Council.

Second, if these ten-dimensional beings reached into our universe and poked a finger into our homes, we would see only a sphere of flesh hovering in midair.

Third, if these ten-dimensional beings grabbed someone who was in jail and deposited him elsewhere, we would see that person mysteriously vanish from jail and then suddenly reappear, as if by "magic," somewhere else. In many science fiction novels, a favorite device is the "teleporter," which allows people to be sent across vast distances in the blink of an eye. A more sophisticated teleporter would be a device that would allow someone to leap into a higher dimension and reappear somewhere else.

Visualizing Higher Dimensions

Our minds, which conceptualize objects in three spatial dimensions, cannot fully grasp higher-dimensional objects. Even physicists and mathematicians, who regularly handle higher-dimensional objects in their research, treat these objects with abstract mathematics rather than trying to visualize them. However, given the analogy with the Flatlanders, there are tricks we can use to visualize higher-dimensional geometric objects such as hypercubes.

The concept of a three-dimensional cube would be alien to the Flatlanders. However, there are at least two ways in which we could convey to them the concept of a cube. First, if we were to unravel a hollow cube, we would, of course, unfold a series of six squares, which can be arranged, say, in the shape of a cross. For us, it is obvious that we can simply rewrap these squares into the shape of a

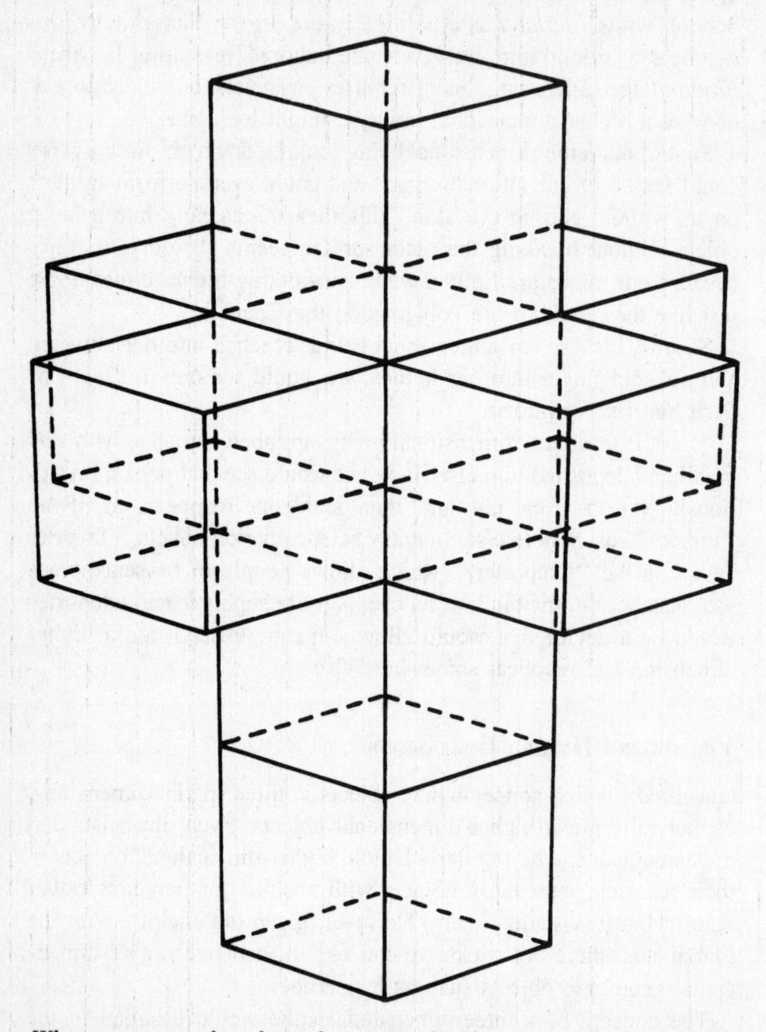

When we unravel a three-dimensional cube, we create a series of squares arranged in a cross. When we unravel a four-dimensional hypercube, we create a series of three-dimensional cubes arranged in a cross-like configuration, called a tesserack (above).

cube. For the Flatlander, this is impossible. Similarly, a higher-dimensional being could convey to us the concept of a hypercube by unraveling it until it becomes a series of three-dimensional cubes, called a tesserack.

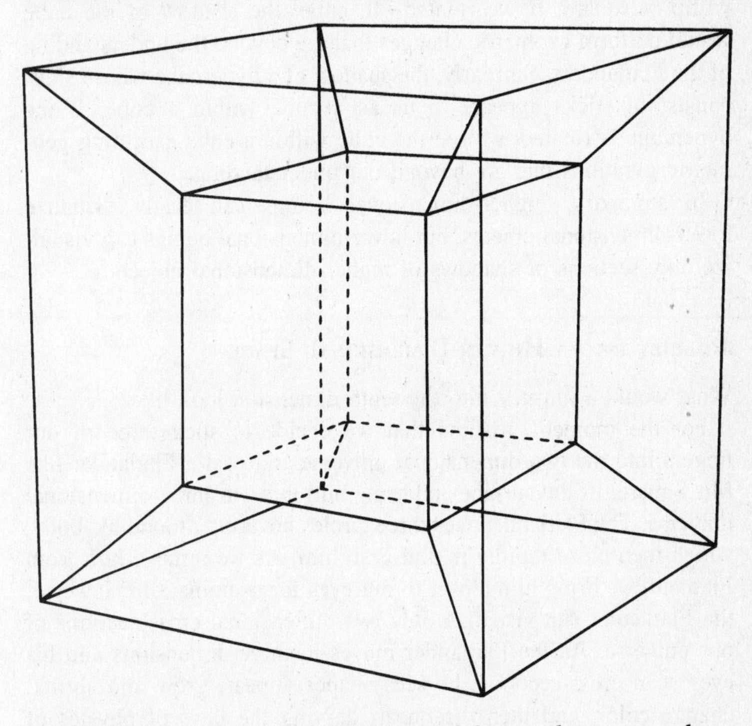

The shadow cast by a four-dimensional hypercube on our three-dimensional universe would look like a cube within a cube.

(Perhaps the most famous illustration of a tesserack is found in Salvador Dali's painting of the crucifixion of Christ, which is on display at the Metropolitan Museum of Art in New York. In the painting Mary Magdalene is looking up at Christ, who is suspended in midair in front of a series of cubes arranged in the shape of a cross. Upon close inspection, one can see that the cross is not a cross at all, but rather an unraveled hypercube.)

There is yet another way in which the concept of a cube could be conveyed to a Flatlander. If the edges of the cube are made of sticks, and the cube is hollow, we could shine a light on the cube and have the shadow fall upon a two-dimensional plane. The Flatlander would immediately recognize the shadow of the cube as being a square within a square. If we rotated the cube, the shadow of the cube would perform geometric changes that are beyond the understanding of the Flatlanders. Similarly, the shadow of a hypercube whose sides consist of sticks appears to us as a cube within a cube. If the hypercube is rotated, we see the cube within a cube executing geometric gyrations that are beyond our understanding.

In summary, higher-dimensional beings can easily visualize lower-dimensional objects, but lower-dimensional beings can visualize only sections or shadows of higher-dimensional objects.

Journey into a Higher-Dimensional Space

What would a journey into the tenth dimension look like?

For the moment, assume that we decide to stick three of our fingers into the two-dimensional universe and peel a Flatlander like Mr. Square off the surface and bring him into our three-dimensional universe. The Flatlander sees three circles hovering around his body, which then close rapidly in and grab him. As we remove him from Flatland, we bring him closer to our eyes for examination. However, the Flatlander can visualize only two-dimensional cross-sections of our universe. As the Flatlander moves in three dimensions and his eyes scan cross-sections, he sees shapes appear, grow and shrink, change color, and then disappear, defying the laws of physics of Flatland.

For example, think of a carrot. We can visualize a carrot in its entirety, but a Flatlander cannot. If a carrot is sliced into many circular pieces, a Flatlander can visualize each slice, but never the entire carrot.

Now let us peel the Flatlander from his two-dimensional universe so that he floats, like a paper doll, in our three-dimensional universe. Drifting in our universe, the Flatlander still cannot visualize a carrot or see where he is going. Because his eyes are on the sides of his face, he can only see sideways, in a two-dimensional plane.

When the tip of the carrot enters his field of vision, the Flatlander will see a small orange circle materialize from nowhere. As the Flatlander continues to drift, he will see the orange circle gradually getting bigger. Of course, the Flatlander is only seeing each successive slice of the carrot, which corresponds to circles.

Then the Flatlander sees the orange circle turn into a green circle (which corresponds to the green carrot top). Suddenly the green circle disappears just as mysteriously as it appeared.

Likewise, if we were to encounter a higher-dimensional being, we might first see three spheres of flesh circling ominously around us, getting closer and closer. As the spheres of flesh grabbed us and flung us into higher-dimensional space, we would see only three dimensional cross-sections of the higher universe. We would see objects appear, change color, grow and shrink in size, and then suddenly disappear. Although we might understand that these various objects actually were part of one higher-dimensional object, we would not be able to visualize this object completely or what life would be like in higher-dimensional space.

The Curvature of Space-Time

What is a space warp?

A space warp is the distortion of the fabric of space-time due to the presence of matter and energy. As we saw in chapter 2, Einstein interpreted this distortion of space-time to be the origin of the gravitational force. To visualize the effect of a space warp, think back to the time of Columbus, when most people thought that the world was flat. To people looking around, the world certainly appeared flat, but only because they were so small compared to the radius of the earth.

Similarly, today we assume that the universe around us is flat, but only because the universe is so large.

If a bug was crawling on the surface of a sphere, it would assume that the sphere was flat, much like the contemporaries of Columbus assumed that the world was flat. However, the bug could travel around the surface of the sphere until it came back to the original point. In this way, we see that a sphere is infinite and unbounded in two dimensions, but finite in three dimensions.

Our universe lives on the skin of this hypersphere, which has been expanding since the Big Bang. Like spots on a balloon that is being blown up, the galaxies are constantly receding from one another. (It's futile to ask, however, where the Big Bang took place. The original expansion of a balloon obviously did not take place anywhere on the surface of the balloon. Similarly, the Big Bang did not take place along the surface of four-dimensional space-time. We need five dimensions to explain where the Big Bang took place.)

In geometry, for example, we learn that the sum of the angles of a triangle is 180 degrees. However, this is true only for a triangle on a flat surface. If a triangle was on the surface of a sphere, the sum of the angles would be greater than 180 degrees. (We say a sphere has positive curvature.) If a triangle was on the inside surface of, say, a trumpet or a saddle, the sum of the angles would be fewer than 180 degrees. (These surfaces have negative curvature.)

Non-Euclidean Geometry

Mathematicians in the past have tried to tell whether our universe is curved. For example, in the nineteenth century the German mathematician Carl Friedrich Gauss had his assistants stand on the tops of three mountains, forming the vertices of a triangle. By measuring angles created by this huge triangle, Gauss tried to determine whether our universe was flat or curved. Unfortunately, he only found that the sum of the angles was 180 degrees, so either the universe was flat or its curvature was too small to be observed.

The mathematics of curved space has a curious history. Around 300 B.C., Euclid of Alexandria, the great Greek geometer, was the first to write down systematically the laws of geometry, starting from a series of fundamental postulates. Over the centuries, the most controversial of these was his "fifth postulate," which simply states that if we have a point and a line, there is only one line that we can draw through the point that is parallel to the original line.

This innocent-sounding, commonsense statement aroused the interest of mathematicians for the next two thousand years, who believed that it was possible to derive the fifth postulate from the previous four. Over the centuries, enterprising young mathematicians periodically announced that they had "proved the fifth postulate,'' but errors were always found in their proofs. Try as they could, mathematicians failed to derive the fifth postulate; in fact, they were beginning to suspect that no proof was possible.

In 1829, the puzzle was solved by the Russian mathematician Nicolai Ivanovitch Lobachevsky. He assumed that it was impossible to prove Euclid's fifth postulate and constructed a new geometry, in which the fifth postulate was actually wrong. This marked the birth of non-Euclidean geometry.

Unfortunately, it was difficult for Lobachevsky to make his work more widely known, because he was extremely poor. Unlike some other mathematicians, he was not a member of the aristocracy or a favorite of the royal court. In fact, he never enjoyed any social position and he often espoused unpopular liberal views, which was risky during the reign of the czars. His isolation was aggravated by the fact that many mathematicians reacted with hostility to the idea that Euclid could have been wrong or incomplete. In fact, Gauss himself had independently come to identical conclusions a few years earlier, but never published his results because of the political backlash it might have created.

In 1854, the German mathematician Bernhard Riemann fully explained this new geometry by showing how to extend these theories to higher dimensions. He demonstrated how all these non-Euclidean geometries could be expressed as geometries on curved surfaces of arbitrary curvature.

Like Lobachevsky, Riemann was not a court favorite. While producing some of the most powerful mathematics of the century, he lived in poverty. To make matters worse, several members of his family depended on him for support. In 1859, his luck finally changed and he landed a professorship in Göttingen. However, years of neglecting his health caught up with him and he died of tuberculosis in 1866, at the age of thirty-nine.

Today Riemannian geometry is the mathematical foundation for general relativity. In fact, Einstein borrowed wholesale large portions of his theory from the mathematicians. Unfortunately, Riemann never lived to see that his theories one day would provide the framework to understand the universe itself.

Where Is the Farthest Star?

Assume, for the sake of argument, that we live on a relatively small hypersphere. What, we ask, is the farthest point in the universe? The ancient philosophers asked this question and wondered what was beyond the farthest object. If the universe was a small enough hypersphere, our telescopes would be able to receive light that traveled completely around it, so we would discover, much to our amazement, that the farthest objects in the universe are the backs of our own heads.

Imagine a bug living on the surface of a balloon. Assume, for the sake of argument, that light can travel only in a circular path along the surface of the balloon. If the bug was peering into a telescope, light from the bug could circulate completely around the balloon until it returned to the bug's telescope. If the bug was peering at the farthest object in the universe, it would eventually see an image of itself peering into a telescope.

Similarly, if we were living in a small hypersphere, light could circulate completely around our universe. Then, through our most powerful telescopes, the farthest object we would see in the universe would be the image of someone (ourselves) peering into a telescope. The farthest star would be Sol, our own sun.

Light, of course, can circulate any number of times around this small hypersphere. This means that if we peer into our telescope again at a slightly different angle, we see an image of ourselves looking at yet another person who is also a carbon copy of ourselves. If we slightly change the angle of our vision, we will see an infinite number of people, each peering into a telescope at the person in front of him. Of course, our eyes are seeing an infinite sequence of people, because our eyes can perceive only three-dimensional objects. In reality, our eyes have only received light which has circled the universe many times.[4]

Black Holes

Although all this seems highly speculative, in 1994 scientists using the Hubble Space Telescope confirmed that the galaxy M87 contained a black hole. In the next few years our space probes will be able to peer into outer space and identify even more black holes, which are the remnants of massive stars that have undergone gravitational collapse.

If we reexamine the picture given to us by Einstein, we see that a black hole is basically represented by a long, trumpetlike depression in the fabric of space-time. However, Einstein noticed years ago that this picture is not entirely correct. It turns out that if there is only one such trumpetlike depression, contradictory results occur. In fact, Einstein was forced to have two such trumpetlike depressions combine in order to give a self-consistent picture of the black hole (shown on next page).

Notice that the black hole appears to be a "gateway" between two entirely different universes. Of course, the gravitational forces would be so great that anyone falling into this black hole would be crushed to death. To Einstein, therefore, it was a mathematical curiosity that these singularities seemed to look like passageways into another parallel universe. For all intents and purposes, gravity becomes infinite at the center of the "bridge" (sometimes called the Einstein-Rosen bridge), and all communication between these two universes is impossible. The very atoms and nuclei of a person would be ripped apart by the gravitational forces at the center.

However, in 1963, physicist Roy P. Kerr discovered that a spinning black hole, instead of collapsing to a point, collapses like a pancake into an infinitely thin ring. Because of the conservation of angular momentum, we expect that most black holes are spinning rapidly, so the Kerr metric, as it is called, is the more appropriate model for a black hole.

The Kerr metric is peculiar, however, because the gravitational forces are not infinite if a person falls directly into the ring perpendicular to its axis. This fact raises the unusual possibility that, in the future, space probes might be sent directly into a rotating black hole and wind up in another, parallel universe. It is possible, in fact, actually to map out the precise paths of such a projectile as it moves from one universe to another.

If we approach this ring from the side, then we will be crushed, just as if we were to try to go through a normal black hole. However,

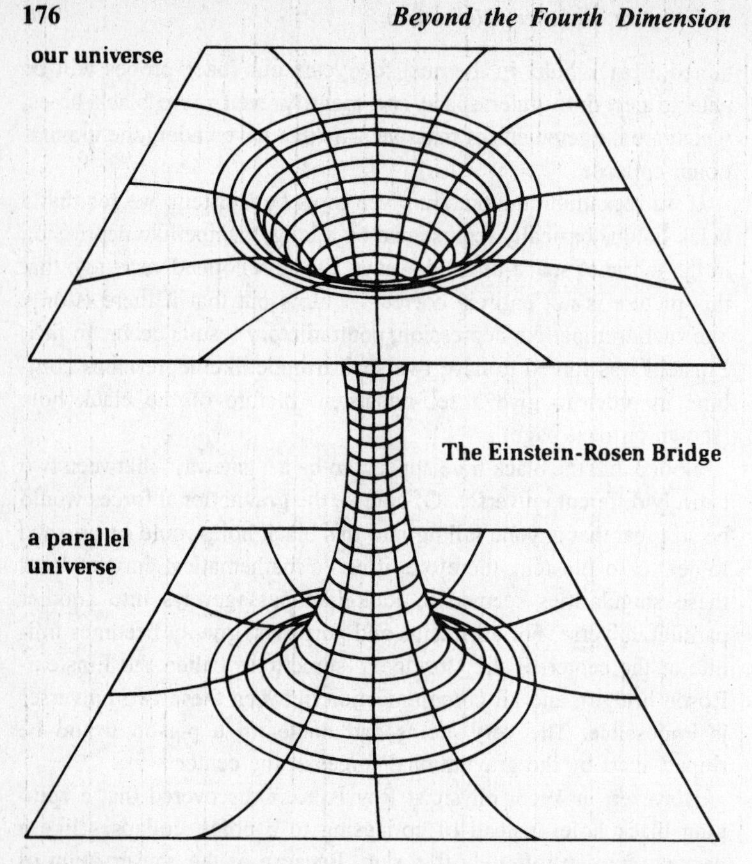

A black hole can be viewed as a gateway to a parallel universe. The "catch" is that the gravitational forces in the middle of the Einstein-Rosen bridge probably make communication between these two universes impossible.

if we approach the ring from the top, the gravitational fields would be enormous, but they would not be infinite.

Stephen Hawking and his colleague Roger Penrose have studied the effects of these strange Kerr black holes. They have found that the neck of the Einstein-Rosen bridge may bend around and come out somewhere else in the universe. This raises the possibility of a dimensional bridge between different parts of the universe.

What might this bridge look like? Imagine, for argument's sake, that we have discovered a Kerr rotating black hole. If we send rockets through the black hole, at right angles to the ring, they would not emerge on the other side of the black hole. In fact, they would emerge on the other side of the universe. In this sense, the bridge could serve as a convenient dimensional passageway to the other side of space.

As fascinating as this possibility is to science fiction writers, it is not clear that these bridges exist. The mere fact that they have been found as solutions to Einstein's equations is not enough. We still must calculate quantum corrections to these worm holes.

Traditionally, quantum corrections have been impossible to calculate in general relativity, so it always has been a matter of speculation whether quantum effects would close the bridge. However, with the coming of the superstring theory, it is only a matter of time before someone calculates what happens in a quantum Einstein-Rosen bridge and determines whether quantum effects seal it off.

Many physicists believe that quantum corrections due to superstrings will seal the entrance, making such journeys into the other side of the universe impossible. If superstring corrections do not seal the dimensional bridge, however, we are left with the interesting possibility of sending rockets directly into a spinning black hole and having them resurface on the other side of the universe.

As strange as these bridges are, even weirder effects of general relativity exist. With the coming of superstrings, we may be able to settle the question of whether bizarre distortions of time are possible.

13

Back to the Future

IN LEWIS CARROLL'S *Through the Looking Glass,* Alice walked through the mirror and entered another universe. In that alternate universe, everything seemed familiar, except there was a twist. In Wonderland, logic and common sense were reversed.

Carroll's real name was Charles Lutwidge Dodgson; he was a mathematician who taught at Oxford and made original contributions to the field of mathematical logic. (Queen Victoria was so enchanted by his children's books that she insisted that he send her his next book. He readily complied, and sent her his latest book on abstract mathematics.)

He originally wrote the Alice in Wonderland series to amuse children with twists of logic. In effect, Carroll was telling children that other worlds were possible with rules completely different from our own.

From the vantage point of modern physics, however, we can ask: What does science say about the possibility of parallel universes similar to our own? What about antimatter universes, mirror universes, time-reversed universes? Surprisingly enough, GUT theories and superstrings say a great deal about the possibility of these different types of universes.

The first person to open the door to the possibility of alternate worlds was Paul Dirac, one of the founders of quantum mechanics, who discovered the theory of antimatter quite by accident.

ANTIMATTER

Dirac was born in 1902, one year after Heisenberg. He was graduated from Bristol University in England at the age of eighteen as an electrical engineer, and could not find a job. He was offered admission to Cambridge University but turned it down because of lack of money. Unemployed, he stayed with his parents and later received a B.A. in applied mathematics in 1923.

In 1925, he heard about the exciting work of Heisenberg, another physicist in his early twenties, who was creating a new theory of matter and radiation: quantum mechanics. Very quickly, and with surprisingly little previous exposure to physics, Dirac plunged ahead and made astonishingly original contributions to the field of quantum mechanics.

In 1928, Dirac, then only twenty-six years old, was bothered by the fact that the Schrödinger equation was nonrelativistic and worked only for velocities much lower than the speed of light. Dirac also noticed that Einstein's famous equation $E = mc^2$ was actually not quite correct. (Einstein realized that the correct version is $E = \pm mc^2$ but did not concern himself with the minus sign because he was creating a theory of forces.) Dirac, because he was creating a new type of equation for the electron (now called the Dirac equation), could not ignore the possibility of matter with negative energy. The minus sign was puzzling because it seemed to predict an entirely new form of matter.

Dirac found that matter with negative energy would look just like ordinary matter but would have the opposite charge. The antielectron, for example, would be positively charged and could, in principle, circle around a negatively charged antiproton, creating antiatoms. These antiatoms, in turn, could combine to create antimolecules and even antiplanets and antistars made of antimatter.

In Dirac's original paper, he was conservative and speculated that perhaps the proton was the counterpart of the electron. However, he left open the distinct possibility that there was a new form of matter predicted by his equation.

The existence of antimatter, first predicted by Dirac, later was demonstrated conclusively by the discovery of the antielectron (dubbed the positron) by Carl Anderson of the California Institute of Technology. Anderson, after analyzing cosmic-ray tracks, noticed in one photograph an electron that was going the wrong way in a magnetic field. Unmistakably, this was an electron with a positive charge.

For his work, Dirac was awarded the Nobel Prize in 1933 at the age of thirty-one and was also awarded the Lucasian professorship at Cambridge, the same position held by Isaac Newton centuries earlier. Anderson won the Nobel Prize soon afterward, in 1936.

So impressed was Heisenberg by Dirac's results that he said, "I think that really the most decisive discovery in connection with the properties or the nature of elementary particles was the discovery of antimatter by Dirac."[1]

When matter and antimatter collide, they neutralize each other and release enormous energy. Examining a large chunk of antimatter would be difficult, if not impossible, because contact with ordinary matter would create a nuclear explosion much larger than that of a hydrogen bomb.

The conversion of matter and antimatter into energy is much more efficient than the release of energy in a hydrogen bomb. In a nuclear detonation, the conversion of matter into energy is only about 1 percent efficient. Antimatter bombs, if they were built, could be up to 100 percent efficient. (Using antimatter to create nuclear bombs, however, is not practical. Although antimatter bombs are theoretically possible, they would be prohibitively expensive.)

Today, detailed experiments are being conducted with antimatter. In fact, at several atom smashers around the world, physicists are producing beams of pure antielectrons that are then made to collide with beams of electrons. (Because the beams are not very intense, the sudden collision of matter and antimatter releases energy but does not cause an explosion.) In the future, the annihilation of matter and antimatter might be useful as a possible energy fuel for space travel (but only if we could find large chunks of antimatter in the universe).

Some people, reading about antimatter in science fiction novels, are surprised to learn that the theory of antimatter is sixty years old.

Perhaps the reason why antimatter's existence is not more widely known is that Dirac was a laconic individual, never one to brag about his achievements. In fact, his taciturn ways were so famous that students at Cambridge University called the "dirac" the unit of volubility. It equaled one word per year.

Going Backward in Time

In the early 1940's, when Feynman was still a graduate student at Princeton University, he introduced another interpretation of the nature of antimatter. In QED, Feynman noticed that antimatter traveling forward in time was indistinguishable from ordinary matter going backward in time.

This discovery allowed a totally new (but equivalent) interpretation of antimatter. For example, if we push an electron with an electric field, it moves, say, to the left. If the electron was going backward in time, it would move to the right. However, an electron moving to the right would appear to us as an electron with positive, not negative, charge. Therefore, an electron moving backward in time is indistinguishable from antimatter moving forward in time. In other words, the electron that Carl Anderson photographed in his cosmic-ray experiments, which acted as if it had a positive charge, was actually going backward in time.

Particles moving backward in time gave a new interpretation of the Feynman diagrams. Assume that we have an electron and an antielectron colliding, releasing a burst of energy. If we reverse the arrow on the antielectron, making it go backward in time, we can reinterpret this diagram. In the new interpretation, one electron goes forward in time, releases a photon of energy, and the same electron goes backward in time.

Feynman, in fact, demonstrated that all the equations of QED were identical whether describing antimatter going forward in time or ordinary matter going backward in time. This bizarre state of affairs makes possible an outlandish theory, proposed by John Wheeler of Princeton University, that the entire universe is made of just one electron.

One day, when Feynman was a student at Princeton, his adviser Wheeler excitedly claimed that he now knew why all electrons in the universe look alike. (Every student of chemistry learns that all electrons are the same. There are no fat electrons, or green electrons, or long electrons.) Wheeler proposed to explain this by assuming that all electrons look the same because they are, indeed, the same electron.

Imagine, for example, the act of creation. Assume that out of the chaos and fire of the Big Bang came only one electron. This electron moves forward in time for billions and billions of years until it arrives at another cataclysmic event—the end of time, or Doomsday. This shattering experience, in turn, reverses the direction of the electron and sends it back in time. When this same electron arrives back at the Big Bang, its direction is reversed once again. The electron is not splitting up into many electrons; it is the same electron that zigzags back and forth like a Ping Pong ball between the Big Bang and Doomsday. Now, anyone sitting between the Big Bang and Doomsday in the twentieth century will notice a large number of electrons and antielectrons. In fact, we can assume that the electron has traveled back and forth enough times to create the sum total of electrons in the universe. (Of course, an object traveling back and forth in *space* cannot create more than one copy of itself. However, an object going back and forth in *time* can indeed have copies of itself. For example, consider the end of the movie *Back to the Future,* when the hero returns to the present just in time to see himself leaving in the time machine. In that scene, there were two images of the hero. In principle, this effect of going backward and forward in time can be repeated an arbitrary number of times, thereby creating an infinite number of carbon copies in the present.)

If this theory is true, it means that the electrons in our bodies are the same electron, the only difference being that my electrons are, say, billions of years older than your electrons. If this theory is correct, it also helps to explain a fundamental principle of chemistry: that all electrons are alike. (A modern-day version of this theory would be to have a one-string universe.)

Can Wheeler's one-electron universe explain the existence of all matter in the universe? Can matter go backward in time and become antimatter? The answer to these questions is formally yes. But no experiment can be performed, according to QED, that can distinguish matter going backward in time from antimatter going forward in time. Therefore, no usable information can be sent backward in time, which eliminates the possibility of time travel. If we see antimatter floating in outer space, it may have reached us from the future but we can't use it to send signals to the past.

Mirror Universes

When Alice looked through the looking glass, she saw a mirror-reversed universe. In that world, most people were left-handed, people's hearts were on the right sides of their bodies, and clocks moved counterclockwise.

As strange as that world might appear, physicists have long thought that such a mirror-reversed universe was physically possible. The equations of Newton, Maxwell, Einstein, and Schrödinger, for example, all remain the same if we reverse them. If our equations make no distinction between left and right, then both universes should be physically possible. This principle, called the "conservation of parity," can be illustrated by a simple example given by Feynman.

Let's say that we have just established radio links with the people of another planet. We cannot see them, but we have deciphered their language and can talk to them by radio. Excited by this extraterrestrial contact, we begin to explain our world to them. We ask, "What do you look like? We have one head, two arms, and two legs." They answer, "We have two tentacles and two heads." They understand everything we say.

Everything is going along smoothly, until we say, ". . . and we have hearts on the left sides of our bodies but not the right."

They reply, "We are puzzled. We understand the meaning of 'heart,' because we have three of them ourselves, but what is the meaning of 'right'?"

This is easy, we say to ourselves. We reply, "You know, 'right' as in 'right hand.' "

Puzzled, they reply, "We understand the concept of a hand, because we have two tentacles ourselves, but which is our right tentacle?"

We think a bit, then reply, "If you rotate your body clockwise, then your body moves to the right."

The aliens reply, "We understand the meaning of rotate, but what is the meaning of 'clockwise'?"

Frustrated, we say, "Do you know the meaning of up and down?"

They reply, "Yes, up means away from the center of our planet, and down means toward the center."

So we add, "When the hands of a clock point up, they will move clockwise, to the right."

They reply, puzzled, "We understand up, we understand clocks, but we still don't understand 'right' or 'clockwise.' "

Exasperated, we make one last try: "If you sit on the north pole, and your planet moves clockwise under your feet, then your planet rotates to the right."

They reply, "We understand the concept of the pole, but how do you tell the North Pole from the South Pole?"

We give up.

The purpose of this story is to show that by radio alone, physicists once thought it was impossible to tell the difference between "left" and "right." Parity conservation, as it was called, which was a cherished notion in physics, states that either a left-handed or right-handed world is a reasonable universe, violating no known principle.

This view of physics came crashing down in 1956 with the work of two young physicists at the Institute for Advanced Study at Princeton. "Frank" Yang, now at SUNY Stony Brook, and another young Chinese emigré, Tsung Dao Lee, now at Columbia, showed that parity was overthrown in the weak interactions. Professor Chien Shiung Wu at Columbia confirmed this experimentally when she showed that cobalt-60 atoms decay by emitting electrons mainly spinning in a preferred direction.

When the results of the experiment were released, physicists were shocked. Pauli, upon hearing the news, said, "God must have made a mistake!"

The world of physics was badly shaken by the theory of Yang and Lee, who showed that it was possible after all to distinguish between left- and right-handed universes. As weird as their theory was, the experimental results were conclusive, and led to their receiving the Nobel Prize in 1957.

Now, armed with the historic results of Lee and Yang, we can get back on the radio and tell the aliens, "I've got it. Get a chunk of cobalt-60, put it in a magnetic field, and the electrons it emits will head for the North Pole. Once you know the meaning of north, you can tell the meaning of clockwise and right.''

The aliens reply, ''We understand what cobalt-60 is. We know which element has sixty protons in its nucleus. We can perform this experiment.''

Thus, with the pioneering work of Lee and Yang, it appears that it is possible to communicate the concept of left and right after all.

Let's say that we have finally built rockets large enough to take us to the alien planet. We agree ahead of time that we shall each shake with our ''right'' hands or tentacles when we meet at that historic occasion.

When the day comes, we meet and stick out our right hands. Suddenly, we notice that the aliens have stuck out their left tentacles.

In a flash, we realize that there's been a mistake. The aliens are actually made of antimatter. All this time, we were speaking to aliens made of antimatter, who conducted an experiment with anticobalt-60 and measured the spins of antielectrons that went south, not north. Then we suddenly get a horrible thought. If we shake the aliens' left tentacles, we will all blow ourselves up in a matter-antimatter collision!

CP Violation

Until the 1960s, it was thought that although parity was overthrown, there was still some hope. A universe that was made of antimatter and that had its left and right hands reversed was still possible. It was believed that the equations of the universe remained the same under a CP reversal (C for ''charge conjugation,'' which turned matter into antimatter, and P for ''parity reversal,'' which interchanged left and right).

Thus, it was still impossible to communicate the concept of left and right to an alien by radio if we did not know ahead of time whether the alien was made of matter or antimatter. Symmetry seemed to be restored to the universe.

However, in 1964, Val L. Fitch and James W. Cronin at the Brookhaven National Laboratory showed that even CP was violated when studying the decay of certain mesons. This meant that the equations of the universe did not remain the same if we reversed matter and antimatter and left and right.

At first, news of CP violation was met with disappointment. It meant that the universe was less symmetrical than originally expected. Although this did not disprove any particular important theory, it meant that nature created a universe much more puzzling than physicists suspected. Today, however, the GUT theory explains why CP violation may actually be a blessing in disguise.

Theories of the origin of the universe have always questioned why we do not see an equal amount of matter and antimatter in the universe. Although it is not easy to differentiate between matter and antimatter in the heavens, astronomers believe that the amount of antimatter in the visible universe is negligible.

What accounts for the imbalance between matter and antimatter? Why should matter dominate in our universe?

Over the decades, highly speculative mechanisms have been proposed hypothesizing that perhaps the matter and antimatter in the universe could be kept apart by some unseen force.

However, the simplest theory comes from the unified field theory. In the GUT and superstring theories, CP is violated. At the beginning of time, there was, as a result of CP violation, a slight imbalance of matter over antimatter (roughly one part in a billion). This means that the matter and antimatter in the universe annihilated each other at the Big Bang, creating radiation, but one-billionth of the original matter was left over. This excess, then, constitutes our physical universe.

In other words, the matter in our bodies is like a fossil, left over from the original annihilation of matter and antimatter at the Big Bang. The reason why matter exists at all is because unified field theories incorporate CP violation. Without CP violation, there is no universe.

Time Travel?

So far, we have discussed only seemingly well-behaved universes that agree with the experimental data. P and CP violations have been measured in the laboratory again and again and are useful in explaining certain features of the early universe.

However, general relativity also allows some universes that are quite difficult to interpret. Some of these universes seem to allow time travel.

When Einstein was alive, each solution to his equations had brilliant success in explaining or predicting aspects of cosmology. For example, the Schwarzschild solution gave us the current description of the black hole. The Nordstrøm-Reissner solution gave us the description of the charged black hole. The Robertson-Walker solution gave us the description of the Big Bang.

However, one solution to this theory raised fundamental questions about the meaning of time itself. For example, in 1949 the Princeton mathematician Kurt Gödel discovered a strange solution to Einstein's equations that was "acausal." (To a physicist, an acausal universe is one in which time is cyclic, repeating itself an infinite number of times, like a motion picture being replayed repeatedly.)

Einstein himself acknowledged the disturbing implications of Gödel's theory. In February 1949, Einstein wrote that Gödel's work was puzzling and raised questions that he could not answer totally. Gödel's solution, he wrote, "constitutes, in my opinion, an important contribution to the general theory of relativity, especially to the analysis of the concept of time. The problem here involved disturbed me already at the time of the building up of the general theory of relativity, without my having succeeded in clarifying it."[2]

Although Einstein could not tear down Gödel's solution, he summed up his criticism by saying that "it would be interesting to weigh whether these are not to be excluded on physical grounds,"[3] that is, whether they could be dismissed because they violated some principle.

In the mid-1960s, University of Pittsburgh physicists E. T. Newman, T. W. J. Unti, and L. A. Tamborini discovered another set of bizarre solutions to Einstein's equations. Their solutions were so weird that they were quickly dubbed NUT solutions, after the originators.

NUT solutions not only allow this strange form of time travel, they permit other strange distortions of space-time. For example, think of walking 360 degrees around a table. Of course, we simply come back to where we started. But now imagine making a

360-degree trip up a spiral staircase. We wind up on the next floor, not at our original starting point.

These NUT solutions allow staircase-type solutions in higher dimensions. This means that if we were to execute a 360-degree trip around a star, for example, we would not wind up where we began, but rather on a different sheet of space-time.

Although Einstein's equations allow for strange distortions of time, we don't have to worry that one day the earth will fall into a NUT solution and come out on the other side of the universe. It is probably not possible, as claimed in *Back to the Future,* to go back in time and have your mother fall in love with you before you are born. These NUT universes, if they even exist, would be beyond our visible universe. Communication with them would be impossible because they are beyond the range of light rays. Therefore, we need not take these solutions of Einstein's equations seriously.

Quantum Corrections to Warped Time

In the 1960s, the Gödel and the NUT universes could be dismissed. It was considered a fluke that Einstein's theories allowed such bizarre universes as solutions.

However, with the advent of the quantum theory, everything became confused. According to the Heisenberg Uncertainty Principle, there is always the chance of making the quantum leap into them, no matter how small the probability. Thus, quantum mechanics reintroduced many of these strange solutions. However, since quantum corrections to Einstein's theory could not be calculated reliably, the entire matter was always a bit of an embarrassment. No definitive statements, one way or the other, could be made.

With the development of the superstring theory, however, the guesswork is eliminated. In principle, all quantum effects are now totally calculable. It will be possible to answer, once and for all, how quantum mechanics does or does not rule out these crazy solutions of Einstein's equations—solutions that admit bridges, falling into other universes, and universes where time travel is possible.

The excitement generated by superstrings, however, is still new and no one has yet calculated these quantum corrections. It would be interesting to see, over the years, how large these quantum corrections are.

Everything from Nothing?

For years physicists have been intrigued by the possibility that the universe came from a quantum transition from nothing (pure space-time, without matter or energy).

The idea of creating something from pure space-time is an old one, dating back to World War II. Physicist George Gamow, in his autobiography, *My World Line,* relates how he first presented this theory to Einstein. Once, while strolling with Einstein on the streets of Princeton, Gamow mentioned an idea proposed by quantum physicist Pascual Jordan. A star, by virtue of its mass, obviously has energy. However, if we calculated the energy locked within its gravitational field, we would find that it is negative. The total energy of the system may, in fact, actually be zero.

What, argued Jordan, would then prevent a quantum transition from the vacuum into a full-blown star? Since the star had zero energy, there was no violation of the conservation of energy when it was created out of nothing. When Gamow mentioned this possibility to Einstein, Gamow recalled, ''Einstein stopped in his tracks and, since we were crossing a street, several cars had to stop to avoid running us down.''[4]

In 1973, Ed Tryon of Hunter College in New York proposed, independently of these early theories about stars, that perhaps the entire universe was created from pure space-time. Again, it appears that empirically the total energy of the universe is close to zero. What if, argued Tryon, the entire universe was created as a ''vacuum fluctuation,'' a random quantum leap from the vacuum into a full-fledged universe?

Physicists pioneering the inflation theory have treated this idea of creating the universe from nothing as a serious concept, however speculative it may be. What relevance does this ''everything from nothing'' theory have for superstrings?

As we saw earlier, the superstring theory predicts that our universe started as a ten-dimensional universe, which was ultimately unstable and collapsed violently down to four dimensions. This cataclysmic event, in turn, created the original Big Bang. However, if the "everything from nothing" theory proves to be correct, it means that perhaps the original ten-dimensional universe started out with zero energy.

At present, superstring theorists are unable to calculate mathematically the precise mechanism by which a ten-dimensional universe can rupture into a four-dimensional one. The mathematics involved is beyond the capabilities of most physicists, because the problem involves a complicated quantum mechanical effect. However, the problem is well-defined mathematically, and hence it is only a matter of time before it is solved. Once the dynamics of how a ten-dimensional universe can crack into a four-dimensional one are understood, we should be able to calculate the energy stored in the original ten-dimensional universe. If the energy of the ten-dimensional universe turns out to be zero, then this would tend to support the "everything from nothing" theory.

Superstrings and Space-Time

Time travel . . . NUTs . . . everything from nothing. These are the outer fringes of the general theory of relativity. Einstein, writing in the 1940s and '50s, could dismiss the more bizarre solutions of his equations with the statement that they might be "excluded on physical grounds." Other skeptical physicists over the years have dismissed these ideas for other reasons, such as that it is impossible to communicate with these weird universes where causality is violated. All this, however, has been a matter of speculation.

Does an acausal universe appear in a quantum theory of gravity? Is the black hole a gateway to another universe? Superstring theory is exciting because it enables us finally to calculate many of the quantum corrections to Einstein's theory and answer these questions once and for all.

All the answers are not yet in, and there is plenty to do in the coming years in superstring research. Perhaps some young readers of this book will become inspired by the quest for the equation of the universe and be the ones to settle some of these questions.

14

Beyond Einstein

WHAT LIES beyond the farthest star? How was the universe created? What happened before the beginning of time? Ever since humans first looked up into the heavens and marveled at the celestial splendor of innumerable stars, we have puzzled over these timeless questions.

At the core of the excitement generated by the superstring theory is the realization that we finally may be closing in on the answers to these questions. It is breathtaking to think that we may be entering an era in which we can provide detailed, numerical answers to the questions posed by the Greeks several thousand years ago.

If the superstring theory is successful, we may be witnessing the culmination of a historic process to which some of the greatest minds in history have contributed. If physicists can show that the superstring theory is a completely finite quantum theory of gravity, then it would be the only candidate for a unified theory of the universe. This would complete the cosmic quest begun by Einstein in the 1930s to unify gravitation with the other known forces.

This, of course, has created tremendous excitement among physicists. Once considered a beautiful but impractical idea, unification has evolved into the dominant theme of theoretical physics in the past twenty years. We may be witnessing the culmination of the past three hundred years of physics, beginning with the work of Newton. As Glashow has said, the isolated strands of physics are now being woven together to produce a tapestry of extraordinary beauty and elegance.

As Schwarz has noted,

> Elementary particle physics is different from all other branches of science in that the problem we're trying to ask is so specific that if you were to completely succeed in answering it, you would be finished. In no other branch of science is there even an abstract possibility of being finished. Chemistry and biology are open-ended. Even other branches of physics, such as condensed matter physics, atomic physics, plasma physics, are open-ended. But in elementary particle theory, you're looking for the fundamental laws, and it's entirely plausible that if this beauty that we're seeking is really there, then there is one concise and beautiful answer that encompasses the whole story.[1]

The implications of this statement are staggering. Historians, for example, consider the discovery of a rare, yellowed manuscript several hundred years old a significant find. Such manuscripts give us an invaluable link to the past, allowing us a glimpse into how people lived and thought many generations ago. Archaeologists think that artifacts unearthed in the ancient ruins of cities several thousand years old are priceless treasures. These artifacts tell us how our ancestors built their cities and conducted their commerce and wars even before written records. Geologists marvel at the beauty of gems that were created deep in the earth's crust hundreds of millions of years ago. Rocks give us invaluable clues to the early earth and help to explain the volcanic forces that shaped the continents. Astronomers, when they probe the heavens with powerful telescopes, are awed by the fact that the light they are receiving was emitted by stars billions of years ago. This ancient light helps astronomers understand what the universe looked like when the stars were still young.

To a physicist, however, the superstring theory allows us to study a time period long before written records, geologic records, or even astronomical records. Incredibly, the superstring theory takes us back to the beginning of time, back to an era when all the forces of the world were perfectly symmetrical and united as one primal superforce. The superstring theory may provide answers to questions about phenomena that are at the center of our existence but beyond all human experience.

Symmetry and Beauty

Astonishingly, we are finding that the universe is a lot simpler than was first expected. In a sense, we are coming full circle. In the time before Newton, scientists believed that the universe was perfectly ordered and structured. By the 1800s, however, with the turmoil leading up to the birth of relativity and quantum mechanics, physics seemed confused and chaotic. Now we seem to be returning to our original idea—an orderly universe—although on a much higher, more sophisticated plane.

The superstring theory shows that symmetry plays a pivotal role in physics. On one hand, we realize that symmetry alone is not sufficient to deduce the laws of physics. But on the other hand, some scientists think that beauty, when based on physical evidence, has been a remarkably accurate guide to theoretical physics. As Schwarz remarked,

> Historically, [beauty has] done very well in theoretical physics when you're probing at the fundamental level. It's probably not a way to do biology, but when you are getting into the structure of fundamental physics at the deepest level, for reasons that nobody seems to understand, the more elegant and simple your scheme is, the more success it seems to have. The whole history of physics over the last two or three hundred years, going back to Newton, shows that very clearly.[2]

We are finding that nature uses more sophisticated but simpler mechanisms than we originally thought to build the universe. Although the mathematics has soared to dizzying heights, the physical picture guiding the mathematics is much simpler than anyone might have expected considering the chaotic data that has spewed forth from our atom smashers.

Furthermore, nature seems much more coherent than before. Previously, in order to get a feeling for the ideas current in modern physics, a layperson would have to read books on black holes, lasers, quarks, quantum mechanics, electromagnetism, and so on. This explosion of information would be confusing to any beginner. Worse, a physics student would have to digest at least twenty volumes to appreciate the current trends in the field. Yet now it is possible to write a book that provides a comprehensive, coherent approach to the entire field, condensing the essential ideas of many volumes into a few visual, pictorial terms. This, in fact, is the underlying theme of this book.

But perhaps the greatest lesson of the past several decades of physics is that nature does not simply find symmetry a convenient feature in building physical structures, nature absolutely demands it. When marrying quantum mechanics and relativity, there are so many pitfalls—a veritable mine field of anomalies, divergences, tachyons (particles going faster than light), ghosts (particles with negative probability), and other diseases—that a tremendous amount of symmetry is necessary to eliminate them.

Simply put, the superstring model "works" because it has the largest amount of symmetry ever found in a physical model. This large set of symmetries, which arises naturally when we write down a theory based on strings rather than points, is sufficient to eliminate these anomalies and divergences.

In some sense, the superstring theory provides one answer to Dirac's objections to the renormalization theory. He could not swallow all the sleight-of-hand tricks invented by Feynman and others to shove the infinities up their sleeve. Dirac found the renormalization theory so artificial and contrived that he refused to believe it could be a fundamental principle of nature. Was Feynman, the prankster and amateur magician, pulling the wool over the eyes of an entire generation of physicists?

The superstring theory provides an answer to Dirac's objections because it requires no renormalization. All the Feynman loop diagrams, physicists believe, are finite due to the enormous set of symmetries inherent in the theory.

Many possible universes that are compatible with relativity can be constructed. Likewise, many universes can be dreamed up that obey the laws of quantum mechanics. However, putting together these two yields so many divergences, anomalies, tachyons, and the like that only one iron-clad solution is probably possible. Some physicists are willing to bet a lot of money that the final solution is superstrings.

Like a Mystery Novel

The tortured development of the unified field theory from its primitive origins to the superstring theory of today resembles in some sense the twists and turns found in a good mystery novel.

Like a mystery novel, the story progresses in stages. In the first stage, the main characters are introduced. This corresponds to the era of Newton, Maxwell, Planck, and Heisenberg, when some of the basic properties of the forces of nature were identified and clarified. This period in physics, however, took an extraordinarily long time, lasting several hundred years, mainly because even the direction of research was not clear. In a murder mystery, by contrast, there is a clear definition of the crime. In physics, only Einstein in the 1930s had a clear vision of the direction physics should take, and he worked in virtual isolation. Moreover, he lacked crucial information on one of the main characters: the nuclear force.

In the second stage, patterns emerge that link various individuals to the crime, giving us the first clues to the culprit's identity. In physics, this corresponds to the confusing but steady progress made in the 1950s and 1960s, where physicists identified SU(3) in the strong interactions and SU(2) in the weak interactions. Lie groups were identified as the proper formalism in which to explain the various forces, but scientists still did not understand their origin or purpose. The quark model was proposed, but there was no understanding of where it came from or what held the quarks together.

In the third stage, definite theories are proposed linking some individuals to the crime, but many false starts and reversals are made. In physics, this corresponds to the era of the 1970s, when gauge symmetry was shown clearly to be the framework for unification of strong, weak, and electromagnetic interactions. There were false starts, however. The S-matrix theory was proposed as an alternative to the quantum field theory, but the S-matrix theory wound up helping to give birth to the string theory. However, the meaning of the string theory was completely misunderstood, and it was discarded during this period.

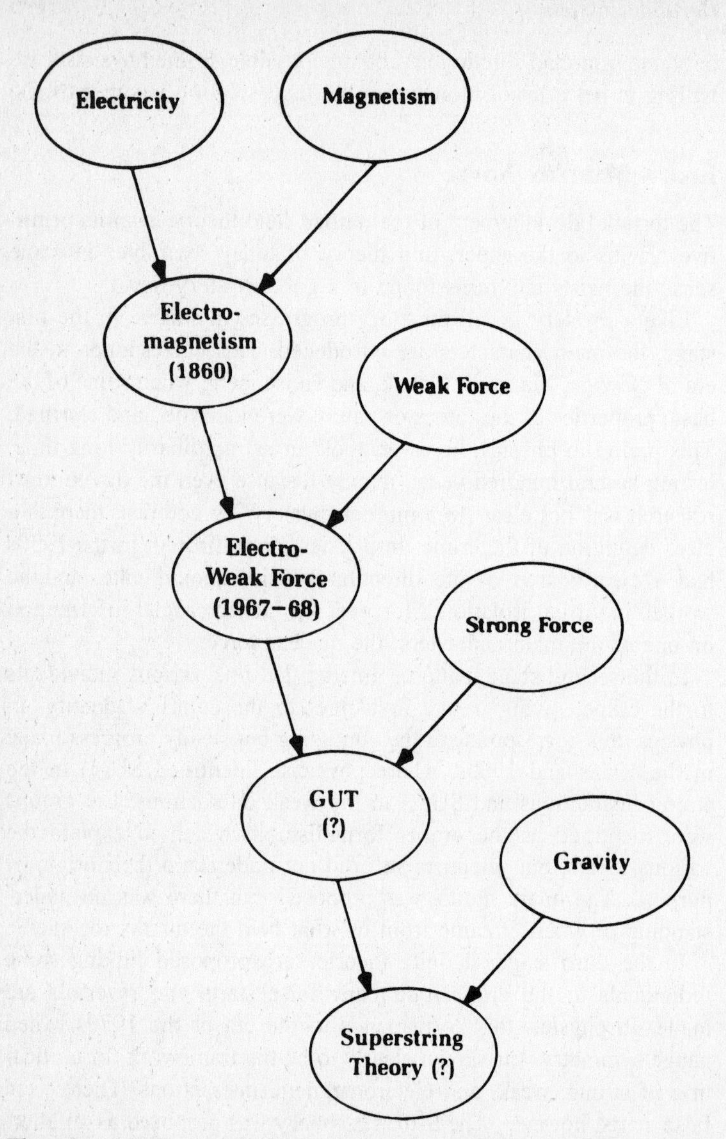

This chart represents the historical sequence of the evolution of the unified field theory, beginning in 1860 with Maxwell's discovery that electricity and magnetism can be united as the electromagnetic force.

In the fourth stage, the clues fall into place and the final conclusion is made. In physics, this corresponds to the last few years, when the superstring theory has emerged as a theory without rival. Although the experimental situation is still up in the air, scientists have enough compelling theoretical results to believe that the superstring theory is the long-sought unified field theory.

Becoming Grandmasters

If the murder mystery has indeed been solved, will physicists eventually put themselves out of a job?

Consider a fable told by Glashow about a visitor from another planet:

> Arthur [is] an intelligent alien from a distant planet who arrives at Washington Square [New York City] and observes two old codgers playing chess. Curious Arthur gives himself two tasks: to learn the rules of the game, and to become a Grandmaster. Elementary-particle physicists resemble the first task. Condensed-matter physicists, knowing full well and with absolute certainty the rules of play, are confronted with the second task. Most of modern science, including chemistry, geology, and biology since the fall of vitalism, is of the second category. It is only in particle physics and cosmology that the rules are only partly known. Both kinds of endeavors are important—one more "relevant," the other more "fundamental." Both represent immense challenges to the human intellect.[3]

As an example, think of a cancer researcher using molecular biology to probe the interior of cell nuclei. If a physicist tells him, correctly, that the fundamental laws governing the atoms in a DNA molecule are completely understood, he will find this information true but useless in the quest to conquer cancer. The cure for cancer involves studying the laws of cell biology, which involve trillions upon trillions of atoms, too large a problem for any modern computer to solve. Quantum mechanics serves to illuminate the larger rules governing molecular chemistry, but it would take a computer too long to solve the Schrödinger equation to make any useful statements about DNA molecules and cancer.

The statement that quantum mechanics solves, in principle, all the problems of chemistry, says everything and it says nothing. It says everything because, indeed, quantum mechanics is the correct language of atomic physics. It says nothing, however, because this knowledge cannot, by itself, cure cancer.

As Glashow says, a unified field theory simply explains to us the rules of the game, but it doesn't teach us how to become Grandmasters.

Thus, the statement that the superstring theory might, in principle, unite all the forces into one coherent theory does not mean the end of physics, but only the opening up of vast new regions of study.

On the Threshold of the Stars

What is remarkable about physics today is that we are making credible statements about the beginning of time while we are, as a species, still young technologically and are just beginning to break free from the imprisonment of the gravity of our planet. We have come a long way intellectually from the time of Giordano Bruno, who was burned at the stake in 1600 by the church for saying that the sun was nothing but a star. But on a technological scale, we are still in our infancy, just beginning to probe the nearest planets in the solar system. Our greatest rockets can barely escape the gravitational pull of the sun.

Yet given our relatively primitive technological development, we still have been able, largely by using the enormous power of symmetry, to make statements about the origin of time itself. On an evolutionary time scale, it has been only perhaps two million years since we left the forests (which is nothing but the blink of an eye) but already we are making careful, rational statements about events that happened billions of years ago, at the beginning of time.

It might be expected that only a more advanced civilization, with vast resources at its disposal, would have been capable of discovering the unified field theory. The astronomer Nikolai Kardashev, for example, has ranked advanced civilizations into three types: Type I civilizations, which control the resources of an entire planet; type II civilizations, which control the resources of a star; and type III civilizations, which control the resources of an entire galaxy.

On this scale, technologically we are still on the threshold of achieving type I status. A true type I civilization would be able to perform feats beyond the scope of present-day technology. For example, a type I civilization could not only predict the weather but actually control it. A type I civilization could make the Sahara desert bloom, harness the power of hurricanes for energy, change the course of rivers, harvest crops from the oceans, and alter the shapes of the continents. A type I civilization would be able to peer into the earth, predict or create earthquakes, and extract rare minerals and oil from inside the earth.

By contrast, at this stage in our development, we can barely control the resources of our nations, let alone an entire planet. However, given the rapid, geometrical explosion of technological development, we can expect to make the transition to a type I civilization and master planetary forces within a few hundred years.

The transition to a type II civilization, which can utilize and manipulate the power of the sun, may take several thousand years, based on the geometric growth of technology. A type II civilization could colonize the solar system and perhaps a few neighboring ones, mine the asteroid belts, and begin to build gigantic machines that can manipulate the greatest energy source in the solar system: the sun. (The energy needs of a type II civilization would be so large that people would have to mine the sun.)

The transition to a type III civilization, which can harness the resources of a galaxy, stretches our imagination to the limit. A type III civilization could master forms of technology that can only be dreamed of now, such as interstellar travel. Perhaps the most revealing glimpse at what a type III civilization might be like can be found in Isaac Asimov's *Foundation* series, which used the entire galaxy as a stage.

Given this perspective, which spans hundreds of thousands of years of technological development, we have made rapid progress in grasping the fundamental laws of nature within just three hundred years of Newton's original theory of gravity.

It is difficult to conceive how our civilization, with its limited resources, eventually will make the transition to a type I civilization and then exploit the full potential of the unified field theory. But Newton and Maxwell, in their lifetimes, probably also never realized that civilization would one day have the resources to send spaceships to the moon or to electrify cities with gigantic electrical plants. In their day, industry and commerce were simply too primitive to absorb or even comprehend the possibilities inherent in their theories.

Fortunately, technological progress proceeds geometrically. However, our brains and imaginations cannot comprehend geometric growth. That is why a science fiction novel, reread decades after it was written, always seems so quaint. In hindsight, we can see that the author's imagination is limited to the technology of his or her time. Science fiction is merely a linear extrapolation or extension of the status quo. That is why science will always be stranger than science fiction.

Given this framework, we can see how difficult it is to predict where the unified field theory will take us, because we are limited by the relative primitiveness of society itself. Even our imaginations are too conservative.

Although we do not have the planetary resources of a type I civilization at our disposal to exploit the practical applications of the unified field theory fully, we certainly have the determination, intelligence, and energy to explore all the theoretical avenues of the unified field theory. Far from the end, it is only the beginning.

Notes

Chapter 1

1. B.M.S., "Anomaly Cancellation Launches Superstring Bandwagon," *Physics Today* (July 1985): 20.

2. M. Mitchell Waldrop, "String as a Theory of Everything," *Science* (September 1985): 1251.

3. Telephone interview, John Schwarz, February 25, 1986.

4. Sheldon Glashow, "Desperately Seeking Superstrings?" *Physics Today* (May 1986).

5. Symposium on anomalies, geometry, and topology, Argonne National Laboratory, Argonne, Illinois, March 29–30, 1985.

6. Freeman Dyson, *Disturbing the Universe* (New York: Harper & Row, 1979), 62.

Chapter 2

1. D. W. Singer, *Giordano Bruno, His Life and Thought* (New York: Abelard-Schuman, 1950), quoted by C. W. Misner, K. S. Thorne, and J. A. Wheeler in *Gravitation* (San Francisco: W. H. Freeman), 755.

2. Abraham Pais, *"Subtle Is the Lord . . ."* (Oxford: Oxford University Press, 1982), 45.

3. Ibid.

4. S. Chandrasekhar, "Einstein and General Relativity: Historical Perspectives," *American Journal of Physics* (March 1979): 216.

5. Pais, *"Subtle Is the Lord . . ."* 462.

6. Ibid., 465.

7. Ibid., 462.

8. E. H. Hutten, quoted by A. P. French, ed., *Einstein: A Centenary Volume* (Cambridge: Harvard University Press, 1979), 254.

9. Letter to H. Weyl, June 6, 1922, quoted by Pais, ''*Subtle Is the Lord . . .*'' 328.

Chapter 3

1. Pais, ''*Subtle Is the Lord . . .*'' 371.

2. In this respect, Planck's constant h plays the same role in the quantum theory that c, the speed of light, plays in relativity. The world of the quantum theory and relativity seems totally alien to us only because the speed of light is such a large, unattainable speed and Planck's constant is so small. In fact, our ''commonsense'' intuition about the universe is correct only if c were actually infinite and h were zero; then both relativistic and quantum effects would disappear exactly.

3. Pais, ''*Subtle Is the Lord . . .*'' 456.

4. Ibid., 13.

5. Max Born and Albert Einstein, *The Born-Einstein Letters* (New York: Walker & Company, 1971), 91.

6. Albert Einstein, Boris Podolsky, and Nathan Rosen, ''Can Quantum-Mechanical Description of Physical Reality Be Considered Complete?'' *Physical Review* 47 (1935): 777ff.

7. Ibid., 461.

8. Ibid., 467.

9. Ibid., 462.

10. Ibid.

11. The majority of physicists resolve the Schrödinger's cat paradox by making a distinction between *microscopic* objects, which are described as strange mixtures of atomic states, and *macroscopic* objects, such as cats. The standard resolution of the paradox assumes that the crucial difference between microscopic events (the collision of two atoms) and macroscopic events (the spreading of cigarette smoke in a room) is that microscopic events are reversible in time, while macroscopic events are not. For example, if we view a film of two atoms colliding, the film looks normal whether it runs forward or backward in time. Thus, at the microscopic level, time can be made to run forward or backward. However, a film of a burning cigarette makes sense only if the smoke is spreading outward, not collapsing back into the cigarette. In other words, the microscopic event, the collision of two atoms, is reversible in time, while the macroscopic event, the spreading of cigarette smoke, is not. Thus, macroscopic events fix the ''arrow of time'' in the direction that increases disorder (the spreading of smoke). Physicists say that the *entropy* of macroscopic events (the measure of disorder) fixes the direction of time and also makes the distinction between reversible microscopic events and irreversible macroscopic events.

The essential feature of making an observation is that it is irreversible—that is, the photographic film can be developed and record the message from photons. The film cannot be "undeveloped." Thus, the transfer of information implies an increase in entropy. Consciousness is not the essential feature of making an observation; machines with no consciousness can make observations. The key feature of observation is the transfer of information, which implies the irreversibility of time. The irreversible transfer of information can be in the form of memory cells in our brains, or photographic film.

12. Gary Zukav, *The Dancing Wu Li Masters* (New York: Bantam Books, 1980), 208.

Chapter 4

1. Richard P. Feynman, "*Surely You're Joking, Mr. Feynman!*" (New York: W. W. Norton, 1985).

2. Dyson, *Disturbing the Universe,* 55–56.

3. John Gribbin, *In Search of Schrödinger's Cat* (New York: Bantam Books, 1984), 259.

4. Heinz Pagels, *The Cosmic Code* (New York: Bantam Books, 1983), 217.

5. Paul Davies, *Superforce* (New York: Simon & Schuster, 1984), 123.

6. Robert P. Crease and Charles C. Mann, "How the Universe Works," *The Atlantic Monthly* (August 1984): 87.

7. Ibid., 89.

8. Sheldon Glashow, Nobel Prize Acceptance Speech, Stockholm, 1979.

9. To obtain an S-matrix, an infinite number of Feynman diagrams must be added. Although at first this seems hopeless, in practice adding just the first set of diagrams in QED rapidly converges to the correct experimental value. This series converges because each set of Feynman diagrams is $1/137$ times smaller than the previous set, so the series gets smaller quite rapidly. This process of adding an infinite set of diagrams that rapidly get smaller and smaller is called *perturbation theory.* The perturbation theory works surprisingly well for QED and the electro-weak interactions, but it fails miserably when applied to the strong and gravitational interactions.

The perturbation theory fails for the strong interactions because the infinite set of Feynman diagrams actually diverges. Terms get larger, not smaller, as we increase the number of loops. Thus, the perturbation theory seems hopeless. To calculate with the strong interactions, we must necessarily abandon the perturbation theory and use *nonperturbative* methods, which are in general quite difficult and often unsolvable. The only method found so far that allows us the chance of calculating the properties of the proton is Ken Wilson's *lattice gauge theory,* which assumes that space-time is defined only on a granular lattice. The lattice gauge theory immediately predicts that the gluon particles condense into stringlike objects that bind the quarks together. This theory requires some of the largest computers ever devised in order to obtain reliable results.

The perturbation theory fails for the gravitational interactions due to an entirely different reason. As Heisenberg noticed decades earlier, each set of Feynman diagrams in quantum gravity has a different dimension and hence cannot be easily added, so Feynman's clever tricks cannot be used. This means that each set of Feynman diagrams, *by itself,* must be finite. Heisenberg thought it would be miraculous if all these billions upon billions of Feynman diagrams were finite all by themselves. In fact, it has now been explicitly checked on computer that a quantum theory of gravity at the two-loop level diverges, thus ending once and for all the hope that quantum gravity can be finite.

Only in the superstring theory does this "miracle" actually happen: Each higher-order diagram is finite by itself and requires no renormalization. The origin of these "miracles" lies in the powerful symmetries built into the superstring theory.

10. Nigel Calder, *The Key to the Universe* (New York: Penguin Books, 1981), 69.

Chapter 5

1. *Scientific American,* July 1994, 26.

2. *Science News,* April 30, 1994, 276.

3. There is yet another theoretical flaw in the GUT theory, which is called the "hierarchy problem." The GUT theory has the curious property that it assumes there is a tremendous gap between two energy scales. The first scale is approximately 10^{16} billion electron volts, which is found only at the beginning of time. The other scale is the energy scale of ordinary particle physics, which is measured in mere billions of electron volts. The GUT theory requires a rigid separation between these two energy scales. (Between present-day energies and 10^{16} billion electron volts, there is a vast "desert" where no new interactions can be found.) However, this rigid separation, which is vital for the theory, begins to collapse once we start to calculate corrections to the theory given by Feynman diagrams. The only satisfactory way of keeping this hierarchy intact when we begin to add

Feynman diagrams is to incorporate four-dimensional supersymmetry into the GUT theory (called SUSY GUT theory).

The SUSY GUT theory, although it solves the hierarchy problem, is quite clumsy-looking. It is hard to believe that anything so contrived can be fundamental. Furthermore, it says nothing about gravity.

From the point of view of a superstring physicist, the problem lies in the fact that the SUSY GUT theory doesn't go far enough. If the SUSY GUT theory is extended so that it becomes the superstring theory, then once again it becomes elegant and simple. As a bonus, it also solves the problem of incorporating quantum gravity.

Chapter 6

1. Unfortunately, Suzuki, hearing that Veneziano had discovered the Beta function independently, never published his own results. Most of the scientific literature, as a consequence, just refers to the "Veneziano model."

2. An earlier, cruder version of the superstring theory, based on strips, was proposed by Leonard Susskind, then at Yeshiva University in New York, and H. B. Nielsen of the Niels Bohr Institute in Copenhagen, as well as by Nambu himself. The strip theory finally was generalized into the complete superstring theory by Nambu (and also, independently, by the late Tetsuo Goto of Nihon University, Kanda, Japan).

3. Laurie M. Brown, "Yoichiro Nambu: The First Forty Years," *Progress of Theoretical Physics* (Kyoto, 1986).

4. Ibid.

5. Dyson, *Disturbing the Universe,* 57.

6. Natalie Angier, "Hanging the Universe on Strings," *Time* (January 13, 1986): 57.

7. Ibid., 56.

Chapter 7

1. Sophus Lie and Elie Cartan showed that there were exactly seven types of Lie groups, which were simply called A, B, C, D, E, F, and G. The first four groups (A, B, C, and D) are labeled by an integer n, which can be arbitrarily large. Thus, there are an infinite number of these groups. However, the Lie groups E, F, and G have intrigued physicists for decades because they allow for a definite number of quarks. Since physicists are always searching for the smallest number of constituents of matter, the groups E, F, and G are likely candidates to describe their symmetries.

The groups A, B, C, and D have historically been the most useful in building models of quarks and leptons. In more familiar notation, we can rewrite these groups as:

$$A(n) = SU(n + 1)$$
$$B(n) = SO(2n + 1)$$
$$C(n) = SP(2n)$$
$$D(n) = SO(2n)$$

where the "S" stands for *special* (the matrix has determinant equal to one), "O" stands for *orthogonal,* "U" stands for *unitary,* and "SP" stands for *symplectic.* Although thousands of papers have been written using these groups to describe elementary particles, the disadvantage is that none of these theories can determine the value of n, which is arbitrary.

However, the groups E, F, and G come only in the following set:

G(2), F(4), E(6), E(7), E(8)

Because there are only a small number of these groups E, F, and G, particle theorists suspect this observation might explain why there are a definite number of quarks. For example, the group E(6) has successfully been used to construct GUT-type theories.

The superstring theory, however, has the symmetry E(8) × E(8), which is more than enough to explain all the known particles and to predict the existence of billions more. When the superstring symmetry is broken, we suspect that it breaks down to E(6), which later breaks down to SU(3) × SU(2) × U(1).

In addition to the original seven groups cataloged by mathematicians, there are also the supersymmetric groups that were originally missed by Lie and Cartan, such as the orthosymplectic group Osp(N/M) and the superunitary group SU(N/M). These latter two groups, in turn, are the underlying symmetry for supergravity and superconformal gravity.

2. Crease and Mann, *The Atlantic Monthly,* 73.

3. Ibid., 75.

4. Ibid.

5. The Yang-Mills theory was also proposed independently by Robert Shaw and by R. Utiyama.

6. Calder, *The Key to the Universe,* 185.

Chapter 8

1. Telephone interview, John Schwarz.

2. Ibid.

3. Technically speaking, the Neveu-Schwarz-Ramond model was not fully supersymmetric when it was first proposed because it contained too many particles. Gervais and Sakita in 1971 proved that the Neveu-Schwarz-Ramond model possessed a two-dimensional supersymmetry on the two-dimensional sheet that the string swept out as it moved in space-time. However, this was not a genuine ten-dimensional space-time supersymmetry.

In 1977, F. Gliozzi, J. Scherk, and D. Olive speculated that the model possessed true ten-dimensional supersymmetry if a subsector of the theory was utilized (the "even G-parity sector"). They used a powerful but obscure mathematical identity (first written down by Carl G. J. Jacobi in 1829!) to show that the bosonic sector and fermionic sector had precisely an equal number of particles if this truncation was made. This conjecture was finally proven by Michael Green and John Schwarz in 1980. Finally, in 1983, Green and Schwarz found the first quantized superstring equation, a supersymmetric version of the Nambu theory. This rigorously proved that the theory was supersymmetric in ten dimensions.

4. Telephone interview, John Schwarz.

5. Michio Kaku and Joel Scherk, "Divergence of the Two Loop Veneziano Amplitude," *Physical Review* (1971): 430, 2000.

6. A version of supergravity also was discovered almost simultaneously by Bruno Zumino and Stanley Deser working at CERN. It should be noted that a more complicated supergravity theory, which possessed far too many particles, was proposed by Richard Arnowitt and Pran Nath of Northeastern University even before the Stony Brook group proposed their theory.

7. Originally, Michael Green and John Schwarz proposed a superstring theory based on the Lie group O(32), which contained both open and closed strings. However, although the O(32) superstring did not have anomalies, the theory had difficulty explaining the experimental features of the known elementary particles. A rival superstring was soon proposed by the Princeton group, based on the Lie group $E(8) \times E(8)$, which contained only closed strings and did not have this experimental problem. Hence, the Princeton superstring, often called the "heterotic" string, is experimentally preferred over the O(32) strings. Technically speaking, when physicists now refer to the superstring, they actually mean the heterotic superstring.

8. Crease and Mann, *The Atlantic Monthly,* 91.

9. Ibid., 91–92.

10. Telephone interview, John Schwarz.

Chapter 9

1. Heinz Pagels, *Perfect Symmetry* (New York: Simon & Schuster, 1986), 209.

2. Dennis Overbye, "Wizard of Time and Space," *Omni* (February 1979): 46.

3. Ibid., 104.

4. Ibid.

Chapter 10

1. Marcia Bartusiak, *Discover* (October 1990): 89.

2. Alan Lightman and Roberta Brawer, *Origins: The Lives and Worlds of Modern Cosmologists* (Cambridge: Harvard University Press, 1990), 305.

3. Ibid., 288.

4. Bartusiak, *Discover,* 90.

5. Lightman, *Origins,* 305.

Chapter 12

1. Pais, *"Subtle Is the Lord . . ."* 330.

2. Ibid.

3. Technically speaking, under quite general physical assumptions, the six-dimensional manifold has a mathematical structure of a "Calabi-Yau manifold." Unfortunately, a straightforward calculation of the breakdown of a ten-dimensional universe into four- and six-dimensional universes is complicated by the complex mathematical structure of these Calabi-Yau spaces. Ultimately, physicists may have to use a nonperturbative calculation on these Calabi-Yau spaces to explain fully why a ten-dimensional universe should break down into four- and six-dimensional universes. The goal is to show that the original ten-dimensional space-time was unstable and "tunneled" via quantum mechanics into a more stable configuration given by the Calabi-Yau manifold of six dimensions and the usual Minkowski manifold of four dimensions.

(It also has been conjectured that the topological structure of these Calabi-Yau spaces eventually will solve the problem of why there are at least three families of leptons and quarks.)

4. At first, it might be suspected that this effect is identical to the illusion created by placing two mirrors next to each other. However, the infinite sequence of images created by two mirrors is strictly imaginary; if we reach out and try to grab one of these images, we only collide with the mirrors. These images exist only because light is reflected back and forth between the two mirrors.

By contrast, to your eye, the infinite sequence of objects in front of you is made of real flesh and blood. It is possible to reach out and grab the image in front of you, which corresponds to your hand encircling the universe and grabbing your own shoulder from behind, much like a dog tries to bite its own tail. However, the brain perceives this effect as an infinite sequence of yourselves lined up in a straight row only because it cannot visualize curved space; it can only interpret the light that falls on the eye.

Chapter 13

1. Calder, *The Key to the Universe,* 25.

2. Schilpp, *Albert Einstein: Philosopher-Scientist,* 687.

3. Ibid.

4. George Gamow, *My World Line,* quoted by John Gribbin, *In Search of the Big Bang* (New York: Bantam Books, 1986), 374.

Chapter 14

1. Telephone interview, John Schwarz.

2. Ibid.

3. Sheldon Glashow and Leon Lederman, ''The SSC: A Machine for the Nineties,'' *Physics Today* (March 1985): 32.

Bibliography

Abbott, Edwin A. *Flatland.* New York: Signet, 1984.

Bernstein, Jeremy. *Science Observed.* New York: Basic Books, 1982.

Calder, Nigel. *The Key to the Universe.* New York: Penguin, 1981.

Carnap, Rudolf. *Philosophical Foundations of Physics.* New York: Basic Books, 1966.

Crease, Robert P., and Charles C. Mann. *The Second Creation.* New York: Macmillan, 1986.

Davies, Paul. *Superforce.* New York: Simon & Schuster, 1984.

Dyson, Freeman. *Disturbing the Universe.* New York: Harper & Row, 1979.

Feynman, Richard P. *"Surely You're Joking, Mr. Feynman!"* New York: Bantam Books, 1986.

French, A. P. *Einstein: A Centenary Volume.* Cambridge, Mass.: Harvard University Press, 1979.

Gamow, George, *One, Two, Three . . . Infinity.* New York: Bantam Books, 1961.

Gibilisco, Stan. *Black Holes, Quasars, and Other Mysteries of the Universe.* Blue Ridge Summit, Pa.: Tab Books, 1984.

Gribbin, John. *In Search of the Big Bang.* New York: Bantam Books, 1986.

———. *In Search of Schrödinger's Cat.* New York: Bantam Books, 1984.

———. *Spacewarps.* New York: Delta/Eleanor Friede, 1984.

Guillemin, Victor. *The Story of Quantum Mechanics.* New York: Charles Scribner's Sons, 1968.

Kaufmann, William. *Black Holes and Warped Spacetime.* San Francisco: W. H. Freeman, 1979.

Lightman, Alan, and Brawer, Roberta. *Origins: The Lives and Worlds of Modern Cosmology.* Cambridge: Harvard University Press, 1990.

Misner, Charles W., Thorne, Kip S., and Wheeler, John Archibald. *Gravitation.* San Francisco: W. H. Freeman, 1973.

Pagels, Heinz. *The Cosmic Code.* New York: Bantam Books, 1983.

———. *Perfect Symmetry.* New York: Simon & Schuster, 1986.

Pais, Abraham. *''Subtle Is the Lord . . .''* Oxford: Oxford University Press, 1982.

Silk, Joseph. *The Big Bang.* San Francisco: W. H. Freeman, 1980.

Snow, C. P. *The Physicists.* Boston: Little, Brown & Company, 1981.

Weinberg, Steven. *The First Three Minutes.* New York: Bantam Books, 1984.

Wolf, Fred Alan. *Taking the Quantum Leap.* New York: Harper & Row, 1981.

Zukav, Gary. *The Dancing Wu Li Masters.* New York: Bantam Books, 1980.

For a survey of the technical literature on superstrings, see:

Alessandrini, V., D. Amati, M. Le Bellac, and D. Olive. *Physics Reports* 1C (1971): 269–346.

Frampton, Paul. *Dual Resonance Model.* Reading, Mass: Benjamin, 1974.

Kaku, Michio. *Introduction to Superstrings.* New York: Springer-Verlag, 1988.

———. *Strings, Conformal Fields, and Topology.* New York: Springer-Verlag, 1991.

Mandelstam, Stanley. *Physics Reports* 13C (1974): 259–353.

Rebbi, Claudio. *Physics Reports* 12C (1974): 1–73.

Scherk, Joel. *Reviews of Modern Physics* 47 (January 1975): 123–164.

Schwarz, John. *Superstrings,* vols. 1 and 2. Singapore: World Scientific, 1985.

———. *Physics Reports* 8C (1973): 269–335.

Index

ABOUT THE AUTHORS

DR. MICHIO KAKU is a professor of theoretical physics at the Graduate Center of the City University of New York. He graduated summa cum laude from Harvard and received his Ph.D. from the University of California at Berkeley, and has taught at Princeton University. He is the cofounder of string field theory, has written eight books, and has published more than seventy scientific papers on superstrings, supergravity, and nuclear physics. He is the author of the widely acclaimed *Hyperspace: A Scientific Odyssey Through Parallel Universes, Time Warps, and the 10th Dimension,* which both the *New York Times* and the *Washington Post* selected as one of the best science books of the year.

JENNIFER TRAINER THOMPSON is a writer and the coeditor with Michio Kaku of *Nuclear Power: Both Sides,* which *The Christian Science Monitor* selected as one of the best books of 1982. Her other books include *Hot Licks: Great Recipes for Making and Cooking with Hot Sauces, The Yachting Cookbook* (coauthored with Elizabeth Wheeler), *Trail of Flame, Jump Up and Kiss Me,* and *The Great Hot Sauce Book.* Her articles on science, culture, travel, and food have appeared in many publications, including *The New York Times, Travel & Leisure,* and *Omni.*